Technology, Business and the Market

Fundamental properties of the universe are transformed into scientific understanding, then developed into new technologies, which are applied to create products and services for business, which then ultimately define our models of organization.

Stan Davis

Technology, Business and the Market

From R. and D. to Desirable Products

JOHN S. SHELDRAKE
Imperial College, London, UK

Routledge
Taylor & Francis Group
LONDON AND NEW YORK

First published 2014 by Gower Publishing

Published 2016 by Routledge
2 Park Square, Milton Park, Abingdon, Oxon OX14 4RN
711 Third Avenue, New York, NY 10017, USA

Routledge is an imprint of the Taylor & Francis Group, an informa business

Gower Applied Business Research
Our programme provides leaders, practitioners, scholars and researchers with thought provoking, cutting edge books that combine conceptual insights, interdisciplinary rigour and practical relevance in key areas of business and management.

British Library Cataloguing in Publication Data
A catalogue record for this book is available from the British Library

Library of Congress Cataloging-in-Publication Data
The Library of Congress Cataloging-in-Publication Data has been applied for.

ISBN 13: 978-1-4094-5455-7 (hbk)

CONTENTS

Contents

ACKNOWLEDGEMENTS

This book, albeit brief, has been a long time in the making and I would like to acknowledge the contributions made to the development of my thinking by Dr Patrick Leevers of the Department of Mechanical Engineering and Professor Paul Robinson of the Department of Aeronautics both at Imperial College London and Professor Stephen Haseler, Director of the Global Policy Institute at London Metropolitan University. Naturally, I accept full responsibility for any mistakes. The book is dedicated to my family and my students with thanks for their support.

ABOUT THE AUTHOR

John Sheldrake's particular area of expertise is in the evolution of management ideas and their application in specific, contextual situations. He has published widely in the areas of management theory, labour relations, local government, transport and health. He has been a consultant in local government and the water industry. He has for many years worked in the Faculty of Engineering at Imperial College London where he is currently Principal Teaching Fellow in the Department of Mechanical Engineering and Visiting Senior Lecturer in the Department of Aeronautics. He was previously Director of Undergraduate Studies in the Business School at Imperial College and a Reader in the Department of Politics and Modern History at London Guildhall University, and is a Fellow of the Royal Historical Society, a Freeman of the City of London and a Liveryman of the Worshipful Company of Fuellers. He was previously a Research Associate at the Management College, Henley and is currently a Professorial Research Fellow at the London-based Global Policy Institute.

e-mail: j.sheldrake@imperial.ac.uk

CHAPTER 1

INTRODUCTION

This short book explores the linkages between science, technology, business and the market – from initial innovation to final consumption. It examines the ways in which scientific endeavour prompts the development of new technical possibilities and processes – some constructive and others destructive in their consequences. It also examines the ways in which scientific endeavour is conditioned and even distorted by contextual issues such as finance and fashion. Of course, the word science has various meanings ranging from the modest 'organized body of knowledge' to the heroic 'search for fundamental explanations of how the world works'. It should perhaps be borne in mind that the word science only gradually replaced the term natural philosophy during the early years of the nineteenth century and that the word scientist was only recognized by the *Oxford English Dictionary* in 1840 (Holmes 2009). Equally, the arbitrary compartmentalization between the various aspects of scientific activity only emerged with the passage of time and the natural philosophers of the seventeenth and eighteenth centuries ranged freely and without inhibition across what we now construe as more or less immutable disciplinary boundaries (Jardine 1999). For the purposes of this book I have taken science to be 'systematic knowledge of the physical world obtained through observation and experimentation'. Further, I have taken technology to mean 'the application of scientific knowledge for practical purposes, especially in industry'. As will be seen, the outcomes of scientific research are most often not spectacular breakthroughs which substantially improve human existence but apparently trivial applications which make our razors run smoother or our fries stay crisper. Technological development, by its very nature, often renders earlier technologies redundant. A major side effect of this is the creation and destruction of skills and jobs with sometimes dire social consequences. Rapid technological progress can therefore bring social disruption as well as social improvement. There are many tensions in all of this which generate the need to explore the costs of progress as well as its benefits. There are also numerous ambiguities and contradictions.

The chapters which follow are arranged thematically and, although they contain some measure of chronological order, there is no overarching narrative. Chapter 2 provides an introductory account of the historical origins of modern science in the West, together with some of the theoretical issues involved. A brief account of the contributions of John Locke, Isaac Newton and David Hume is given, together with an examination of the work and influence of two proto-scientists – the potter Josiah Wedgwood and the surgeon John Hunter. Finally, an account of the emergence of modern management during the nineteenth and early twentieth centuries is provided. Chapter 3 traces the linkages between scientific endeavour, applied science and technological progress through the mediation of commercially sponsored research and development (R&D). It also examines some aspects of the philosophy of science through the debate between Karl Popper and Thomas Kuhn. The subject of industrial R&D is considered through an account of the work carried out at the Edison, Bell and RCA laboratories. An account is given of the combination of technological and managerial development which resulted in the emergence of mass production at the Ford Motor Company and the development of the Sloan Structure at General Motors. Finally, the further evolution of modern management during the early years of the twentieth century is examined. Chapter 4 examines the nature of entrepreneurship and innovation, and offers a brief account of the role of venture capital. The work of the economist Joseph Schumpeter is outlined, specifically his idea of creative destruction as the driving force behind capitalism. A consideration of the role of innovation management leads to an analysis of three specific examples – Ray Kroc at McDonald's; Gordon Moore and Robert Noyce at Intel; and Bill Gates at Microsoft. Chapter 5 examines the developing role of the industrial designer and the links between design and product innovation. The influence of designers such as William Morris, Peter Behrens and Raymond Loewy is considered, together with the design policy developed by Eliot Noyes at IBM, Dieter Rams at Braun and Frank Pick at London Transport. Finally, the phenomenon of retro is examined, together with an example in the shape of Harley-Davidson motorcycles. Chapter 6 examines some of the impacts of technological change by reference to the emergence of the quality movement in Japan and the evolution of advanced manufacturing. An account of the origins and development of containerization is provided through a consideration of the inputs of Malcolm McLean and Keith Tantlinger. The impact of containerization in terms of facilitating global trade and outsourcing is examined through the specific example of Steve Jobs and Apple. Finally, an account of the development and managerial impacts of the Internet are given through an account of the ideas of Rosabeth Moss Kanter. Chapter 7 examines the issues of business policy and strategy and links them to some of the key elements in marketing. Examples are provided in the shape of Boeing versus Airbus and the future development of aircraft design; AMEX and improved customer service; and Procter and Gamble and the development of brand management and market research. Chapter 8 examines some of the criticisms of marketing and its role within the prevailing capitalist economic system. An account of the so-called throw-away society is given using insights provided by Vance Packard and Thorstein Veblen with an example in the shape of the Gillette safety razor. The phenomenon of planned obsolescence is examined using the example of General Motors and the work of Charles F. Kettering and

Harley J. Earl. Finally, two Marxist-inspired critiques of the system are provided using insights from Herbert Marcuse and Harry Braverman. Chapter 9 is linked in tone to the previous chapter and examines the topic of sustainability including the challenge of environmentalism, together with the impact of the corporate responsibility and business ethics movement; theoretical insights are provided by an examination of the work of C. P. Snow and Fritz Schumacher. Chapter 10 provides some concluding remarks and pointers to likely future developments. The texts I have used can be found in the extensive bibliography.

CHAPTER 2

CONTEXTUAL AND THEORETICAL CONSIDERATIONS

INTRODUCTION

This chapter provides an introductory account of the rise of modern science in the West, together with some of the theoretical issues involved. We spend our lives in a world dominated by science and technology and it is almost impossible to conceive of civilised existence without the products and services which science and technology provide. A few examples will give the flavour – electricity, gas and the supply of clean water; air travel, automobiles and railways; computers, television, telephones, the Internet; relatively safe surgery, dentistry, the control of (most) epidemic diseases; central heating, refrigeration, plastics and artificial fibres and so on and so on. Perhaps the greatest single change behind all of this was the transition (in eighteenth- and early nineteenth-century Britain) from an *advanced organic* economy based on agriculture, to a *mineral* economy based on coal, iron and associated manufacturing (Wrigley 1988). A central technological feature of this change was the development of the steam engine and its deployment in factories as the prime means of driving machinery and on railways as the means of traction. The development of metallurgy and the design and construction of factories (as well as railways) were largely informed by empirical observation and carried out by practical engineers rather than 'scientists' (Habakkuk 1962). Having said this, however, science as we currently understand it was gaining ever greater potency and emerging into a position of influence alongside engineering. Not least, the insights derived from Newtonian physics were becoming diffused across the wider society as science, education and industry came together to an unprecedented extent during the first half of the nineteenth century (Huff 2011). This combination of scientific thought and empirical application was informed by specific philosophical and theoretical reflections.

A somewhat arbitrary but nevertheless key date in the history of Western science is 1543. In that year Nicolaus Copernicus (1473–1543) published *On the Revolutions of the Heavenly Spheres*, demonstrating that the Earth, like the other planets, revolves around the Sun while Andreas Vesalius (1514–64) produced *On the Fabric of the Human Body*, the first modern study of anatomy. Essentially, Copernicus provided a fresh insight into what exists *outside* of us in space, while

Vesalius did the same for what happens *inside* of us in our bodies (Trombley 2011). Both of these thinkers had embarked on major scientific voyages of discovery characteristic of what became known as the Scientific Revolution. In the early years of the seventeenth century the statesman and philosopher Francis Bacon (1561–1626) outlined the possibilities for technological progress and also advocated a scientific method based on careful observation and the amassing of data – the method known as induction – moving from the particular to the general (Ball 2013). This was the approach adopted by Bacon's contemporary, William Harvey (1578–1657) whose discovery of the circulation of the blood was published in his book *On the Motion of the Heart and Blood* in 1628. Bacon's ideas influenced the founding of the Royal Society in London in 1660 under the patronage of Charles II (1630–85) with the purpose of fostering understanding of the natural world through observation and experiment and having as its motto *nullius in verba* – take nobody's word for it. In the sections of this chapter which follow, the contributions of John Locke (1632–1704), Isaac Newton (1642–1727) and David Hume (1711–76) will be considered, together with work of two proto-scientists – the potter Josiah Wedgwood (1730–95) and the surgeon John Hunter (1728–93). Finally, a brief account of the origins of modern management is given.

THE IMPACT OF LOCKE AND NEWTON

Broadly speaking, three revolutions occurred or were initiated in Britain during the seventeenth and eighteenth centuries. First, a *political revolution*, which saw absolute monarchy overthrown and replaced, initially by a republic, and subsequently by a constitutional monarchy with power shared between the King and Parliament. Secondly, a *scientific revolution*, which generated the emergence of a new way of understanding the universe and humanity's position in it and in the process challenging beliefs which had in some cases prevailed for hundreds of years. Finally, an *industrial revolution* began which gathered pace during the nineteenth century and, in the course of a century and a half, changed Britain decisively from a sparsely populated, rather backward agricultural country into a densely populated, urbanized industrial one with a burgeoning overseas empire. Perhaps it is more accurate to describe what occurred as a first Industrial Revolution, associated with the development of the steam engine and the application of factory discipline. This was followed by a second industrial revolution, dating from around 1840 to the 1950s, which was characterized by a radical transformation in transportation and communication through the development of railways, the telephone, radio, automobiles and aeroplanes. Finally, a third industrial revolution began in the 1950s based on information technology and placing greater emphasis on the 'service sector' and science-based industries such as synthetic chemicals and pharmaceuticals, biotechnology, electronics and computer hardware and software (McCraw 2000).

The political and social revolutions which occurred in the seventeenth century were accompanied by the emergence of new financial arrangements and the shift in social relationships associated with capitalism. London was at the centre of these developments, becoming the largest city in Europe and, together with commerce, providing a focus for intellectual inquiry and the arts. In terms of intellectual inquiry two specific thinkers of the period had an immense and lasting impact: John Locke and Isaac Newton. Locke, as well as being an influential political theorist and ideologist for the so-called Glorious Revolution of 1689, was also a physician and philosopher. Locke's political thought, contained in his *Two Treatises of Government* published in 1689, advocated limited government, opposing the notion of 'the Divine Right of Kings' and also his older contemporary Thomas Hobbes' (1588–1679) arguments for political absolutism. For Locke, human understanding is inevitably limited and this limitation on absolute knowledge prompted him in the direction of qualified toleration of the beliefs of others – a sort of proto-liberalism pointing forward to the ideas of John Stuart Mill (1806–73) a century and a half later. All of this was consistent with Locke's philosophical position which he outlined in *An Essay Concerning Human Understanding* published in 1690. Locke was an empiricist and held the belief that all knowledge is ultimately derived from experience. On this view there is nothing which can be known to be true or false independent of experience. According to Locke the human mind at birth is like a blank sheet of paper, a *tabula rasa*, on which experience will, so to speak, 'write'. What can be known with any certainty is therefore restricted to that which the senses can perceive. There are obvious problems with this, not least the extent to which we can trust our senses, but for Locke's purposes it enabled him to develop an approach which describes knowledge as merely the aggregate of things which reach our senses. Basically, the world is there to be discovered using our five senses augmented and enhanced from time to time by instrumentation and techniques of various kinds – in Locke's period the development of the telescope and the (primitive) microscope are obvious examples. Locke's views, however limited, left a lasting impression on philosophy in the Anglo-Saxon world – not least a suspicion of overarching philosophical systems on the Continental pattern. Interestingly, they also had a substantial impact on the so-called Enlightenment in eighteenth-century France and in the emerging United States of America – in both cases contributing to the ferment of ideas leading to political revolution (Himmelfarb 2008).

Isaac Newton, like Locke, was a polymath who carried out original research in optics, mathematics, mechanics and gravitation as well as producing historical studies and investigations in chemistry and alchemy. Newton considered his most productive years to be the mid-1660s, a period when he spent much of his time away from his work in Cambridge (which was threatened by bubonic plague) at his family home in Lincolnshire. It was there, according to the well-known story (or probably myth), that he saw an apple fall from a tree and asked himself the question 'if the apple on the tree was at rest, then it gradually accelerated while falling, what causes that acceleration?' His answer, of course, was gravity. Newton developed and refined his initial insight, partly in correspondence with Robert Hooke (1635–1703), the inaugural curator of experiments at the Royal Society, and eventually produced a revolutionary synthesis of astronomy, mechanics and mathematics in his *Mathematical Principals of Natural Philosophy* (often referred to as the *Principia*) published in Latin in 1687 with revised editions in 1713 and 1726. An English translation appeared in 1729. Newton's work is ranked as one of the greatest achievements in abstract thought and became the dominant scientific *paradigm* (see Chapter 3) until it came under increasing pressure in the final years of the nineteenth century (Okasha 2002).

DAVID HUME AND THE PROBLEM OF INDUCTION

David Hume (1711–76) was a philosopher and historian who both built on and challenged the empiricist philosophical foundations provided by Locke and, as perhaps befits an historian, he was a sceptic. Hume was a 'man of letters' who published widely, his most significant works for our purposes being *A Treatise on Human Nature* (1739) and *Enquiry Concerning Human Understanding* (1748). In these books he advanced a theory of human knowledge (epistemology) in which he claimed that human beings are basically creatures of instinct and habit whose mental lives are dominated by passion rather than reason and whose beliefs are formed by mechanisms of association and custom rather than a priori reflection. On this view reason is the servant of the passions rather than the other way around. What we experience as the causal regularity of the world is merely a product of custom. We experience certain natural sequences and this experience leads us to anticipate the ongoing repetition of these sequences. A familiar cause is associated in the mind with the usual effect and from this we construct a set of rules calculated to predict natural sequences and expectations (Mautner 2005). This brings us to the problem of induction – sometimes referred to as 'Hume's problem' – briefly put it is that no sequence of events, or number of confirming observations, allows us to say with certainty that such events will always occur in the future or that all objects of the class observed will always conform to what we have so far observed. In the customary example, if every swan we have so far observed is white, we cannot say with certainty that 'all swans are white'. However, if we observe a single black swan it would permit us to say that 'not all swans are white'. Although we can introduce levels of probability to take account of our lack of certainty, we nevertheless have to admit to the limitations of our knowledge which therefore always remains tentative and subject to change. This leads us on to the work of Karl Popper which we will examine in Chapter 3.

JOSIAH WEDGWOOD – POTTER, INDUSTRIALIST AND EMPIRICAL GENIUS

Josiah Wedgwood was born in Burslem, Stoke-on-Trent in 1730 into a family of potters who were also religious dissenters and therefore excluded from contemporary public life. He was the youngest of 13 children and, owing to the death of his father, left school when he was only nine and became apprenticed to an elder brother. At the age of eleven he suffered an attack of smallpox which left him with a permanently weakened right leg and unable to operate the treadle of the potter's wheel. Instead, he spent his time researching the craft of pottery and working on new designs. During the 1750s, having completed his apprenticeship, Wedgwood entered into partnership with perhaps the leading pottery maker of the day, Thomas Whieldon (1719–95). Over the next decade, Wedgwood continued his experimental work and eventually developed a lead-glazed, cream-coloured earthenware known as 'creamware' which he patented in 1763. Wedgwood ended his partnership with Whieldon in 1759 and set up in business on his own account producing his pottery for a rapidly growing market, including among his customers Queen Charlotte (1744–1818), the wife of George III (1738–1820). Royal patronage massively expanded the demand for Wedgwood's ceramics, not least among the commercial classes whose new wealth was supporting an emerging 'consumer boom' (Wilson 2012). In 1764 Wedgwood married his wealthy third cousin Sarah Wedgwood (1734–1815). As well as being intelligent, creative and generally accomplished, Sarah also had money – money which Josiah Wedgwood was able to invest in the development of pottery manufacture on an industrial basis at the factory he established in Burslem. Here, and later at his new works called Etruria, Wedgwood combined innovative factory layout with the advanced division of labour, subdividing the skills of the potter (i.e. mixing, shaping, decorating, glazing and firing) in order to maximize efficiency.

The expansion of Wedgwood's business was inhibited by the contemporary transport system; based on packhorses and horse-drawn waggons it was both slow and likely to result in many breakages of his fragile wares. In search of a remedy to this problem Wedgwood became involved in Britain's first transport revolution – the engineering of the canals. During the 1760s he collaborated with the canal pioneer Francis Egerton (1736–1803) the Third Duke of Bridgewater, and the engineer James Brindley (1716–72), in the creation of the Trent and Mersey Canal, linking the River Mersey near Runcorn in Cheshire to the River Trent in Derbyshire – a distance of over 90 miles and one of the civil engineering marvels of the day. Wedgwood's campaign for the building of the Trent and Mersey Canal brought him into contact with the physician Erasmus Darwin (1731–1802). Darwin had studied at St John's College, Cambridge and at the Edinburgh University Medical School and was associated with the so-called Lunar Society of Birmingham. The Lunar Society was an informal learned society of industrialists and scientists, or natural philosophers as they were then called, which met and also corresponded from the 1760s through to the early years of the nineteenth century. The name Lunar Society arose from the custom of members gathering at the time of the full moon to make travelling easier and safer at a time when street lighting was minimal. Although the actual composition of the group varied, it included such important figures as the inventor and mechanical engineer James Watt (1736–1819) and his business partner, the manufacturer Matthew Boulton (1728–1809), who between them revolutionized the development and application of the steam engine; the chemist Joseph Priestley (1733–1804) who discovered oxygen; and the botanist William Withering (1741–99) who discovered digitalis (Uglow 2002). Interestingly, the naturalist Charles Darwin (1809–82) who outlined the theory of evolution in his book *On the Origin of Species*, published in 1859, was a grandson of Erasmus Darwin on his father's side and of Josiah Wedgwood on his mother's side. Through the Lunar Society, Wedgwood was networked with the latest in contemporary engineering and scientific thinking and he applied a similarly scientific approach to his business. As well as being active in advanced design, technical and management issues, he was also a pioneer in marketing and branding, opening showrooms for his products in London and also the fashionable spa town of Bath. Wedgwood was an enlightened employer, pioneering the building of model housing for his workers and actively campaigning against slavery. Having said all of this, he was nevertheless constrained to suffer the limitations of his era. When his damaged right leg had eventually to be amputated in April 1768 the operation was carried out by the local surgeon in Wedgwood's house without the benefit of anaesthesia – which leads us to our next topic.

JOHN HUNTER AND THE RISE OF MODERN SURGERY

The rise of modern surgery displays a clearly evolutionary, piecemeal process of development – a fortuitous coming together of science, technology and technique. For centuries surgery was the poor relation of medicine, concerned with low-status interventions relating to the extremities of the human body. Among the procedures carried out by surgeons 'cupping' and 'bleeding' formed an important element and, for many centuries, established an association between barbers and surgeons. Not until the mid-eighteenth century did the surgeons in England break away to form the Company of Surgeons in London. Cupping and bleeding were based on the theory of the humours – the idea that the human body is in some way organized around the balance of four elements and that disease is caused by an imbalance between the four. If an imbalance occurs then interventions must be made to restore the balance – hence the removal of blood to reduce the level of a fever. Medical leeches were used for the same purpose. Even as medical science began to refute the theory of the humours, cupping, bleeding and the use of leeches persisted and elements of the system can still be found today in various traditions of medicine and psychology (Magner 1992). Surgery suffered from the stigma of shedding human blood and was limited in its scope by an acute lack of knowledge of the human body. Until the time of Andreas Vesalius medical understanding was based on the work of Galen of Pergamon (129–200) who had dissected pigs and monkeys but not human beings. It was only during the seventeenth century that Galen's ideas began to be seriously challenged, specifically by William Harvey's discovery of the circulation of the blood. However, even as a more rational, scientific approach to medicine was emerging, the stigma associated with anatomy persisted. In England the only legal source of cadavers was the bodies of executed criminals, condemned not just to death but to being publicly dismembered at Barber Surgeons Hall – a situation not fully remedied until the passing of the Anatomy Act in 1832. The individual generally credited with originating the rise of surgery from an empiric craft to something resembling a science-based profession is John Hunter (Moore 2005).

Hunter was born on a farm on the outskirts of Glasgow, the youngest of ten children and the younger brother of the successful anatomist William Hunter (1718–93). William Hunter had studied medicine in Scotland before moving to London to train in obstetrics, becoming physician to Queen Charlotte, and founding a private medical school in Great Windmill Street, Soho. Meanwhile, John Hunter had left school when he was 13 and started to learn the trade of cabinet maker. In 1748 he visited William Hunter in London and began to work as his assistant in carrying out dissections (as well as procuring bodies for the purpose) before undertaking the preliminary training to become a surgeon, being appointed assistant surgeon at St George's Hospital, Hyde Park in 1756 and full surgeon in 1768. In addition to his surgical and teaching activities, John Hunter also carried out research in comparative anatomy, amassing a vast collection of specimens, some of which are displayed in the Hunterian Museum at the Royal College of Surgeons in London. Hunter favoured direct observation of nature in the empiricist tradition of Locke and attempted to meaningfully classify the materials he examined – essentially taking a scientific approach and even, in the case of his researches into venereal diseases, undertaking self-experimentation. He experimented with the transplantation of teeth, publishing his findings during the 1770s in two parts as *The Natural History of the Human Teeth* and *A Practical Treatise on the Diseases of the Teeth* (Moore 2005). However, it was in the development of surgical techniques that John Hunter had perhaps his greatest influence, not least in teaching them to numerous pupils and generating a sort of surgical apostolic succession at St George's Hospital through the work of his brother-in-law Sir Everard Home (1756–1832) and Home's pupil Sir Benjamin Collins Brodie (1783–1862). Hunter's surgical interventions were inevitably limited by the absence of anaesthesia and knowledge of antiseptics – like Wedgwood he was constrained to live in his era.

The introduction of anaesthesia in the 1840s gave surgeons more time and increasingly complex operations were undertaken. The surgeons now began to rise in status compared to the physicians – the Company of Surgeons having become the Royal College of Surgeons in 1800. Improvements in the quality of microscopes, together with a growing awareness of the significance of tissues, led to advances in physiology and a greater understanding of the human body. Nevertheless, in a period when there was no understanding of the causes of infection, surgery remained a risky business. Not until the introduction of antiseptics in the 1860s did surgery become relatively safe. The fact that surgeons now began to sterilize their instruments led to rapid changes in the construction of the instruments themselves and the materials from which they were constructed. The problem of surgeons working blind was at least partly solved by the end of the nineteenth century by the discovery by Wilhelm Conrad Rontgen's (1845–1923) discovery of so-called X-Rays. By the early twentieth century surgery had become an honourable, prestigious and lucrative profession and 'celebrity' surgeons had emerged, such as Sir Frederick Treves (1853–1923) who discovered John Merrick (1862–90) the so-called Elephant Man, operated on the future Edward VII (1841–1910) for appendicitis and led a medical mission to the South African War.

THE EMERGENCE OF MODERN MANAGEMENT

Management defined in terms of the marshalling and organizing of human and material resources has existed for thousands of years. However, as we have seen in the example of Josiah Wedgwood, the technical and social changes generated by the industrial revolution radically shifted attitudes regarding such issues as the division of labour, the design and utilization of machinery, anticipated levels of production, returns on capital employed, the significance of time, the control of the labour process and the design of products. In his book *The Wealth of Nations* (1776) the Scottish moral philosopher and associate of David Hume, Adam Smith (1723–90), provided an extensive account of the improvements in productivity which could be achieved as a result of the division of labour. Smith saw the division of labour as the key to economic growth and also the means of escaping scarcity and sustaining population growth. In his plan for the Panopticon penitentiary of 1791, the utilitarian philosopher Jeremy Bentham (1748–1832) put forward a design for a prison in which the inmates would be kept in a state of constant inspection, or what we would now term surveillance. The Panopticon was never constructed but the plan strongly influenced the design of prisons, asylums, workhouses, hospitals and, above all perhaps, factories. Although the factory system never fully dominated manufacturing, its advantages in terms of efficiency and control of labour were self-evident to thinkers such as the inventor and mathematician, Charles Babbage (1791–1871) and the Scottish medic and chemist Andrew Ure (1778–1857). In *On the Economy of Machinery and Manufactures*, published in 1832, Babbage analysed the operation of the factory system and what would now be called labour relations. In his *The Philosophy of Manufactures*, published in 1835, Ure studied the impact of the enhancement of human and animal effort by machines and advocated the replacement of skilled workers with semi-skilled and unskilled operatives. The process of replacing skilled workers with machines was paralleled by experiments in the bulk production of interchangeable parts which enabled workers to assemble the finished products quickly and in large quantities.

In 1803 a series of machines for manufacturing pulley-blocks was installed at the Royal Naval dockyard at Portsmouth. These machines, designed by the French expatriate civil engineer Marc Brunel (1769–1849) and constructed by the machine-tool maker Henry Maudslay (1771–1831), provided perhaps the first example of mass production using all-metal machine tools. In 1812 the American inventor Eli Whitney (1765–1825) took a similar approach to the manufacture of rifles. What became known as the American system of manufacturing placed emphasis on analysing mechanisms, breaking them down into interchangeable constituent parts and then designing them for mass production (Sheldrake 2003). During the nineteenth century these techniques were applied progressively to the production of guns, clocks, bicycles, sewing machines and, most importantly, automobiles. Although Britain led the world in manufacturing during the mid-nineteenth century it began to lose its edge as the century progressed and competitors such as France, Germany and the USA underwent industrialization. In 1880 there were approximately 2,700,000 workers employed in manufacturing in the USA. By 1900 the figure had risen to 4,500,000 and by 1920 it had reached 8,400,000. New systems of manufacturing prompted American business to invest in ever-larger production units. Although steel plants and locomotive works initially led the field it was automobile manufacture which ultimately yielded the largest plants. For example, at his Highland Park Plant in Detroit, Michigan, Henry Ford (1863–1947) employed almost 13,000 workers in 1914 and 33,000 in 1916. When his River Rouge Plant opened in 1928, 100,000 workers were employed in what became the largest integrated factory in the world – we will examine Ford's production techniques in Chapter 3. The rapid development of industrialization in the USA, together with increasing levels of technological complexity, stimulated the emergence of professional managers, particularly in mechanical engineering. By the 1880s the American Society of Mechanical Engineers had already become the focus of discussions on the development of techniques for the efficient management of industry. It was in this context that Frederick Winslow Taylor (1856–1915) formulated the ideas which eventually became known as scientific management. As with the more generalized development of science, Taylor's ambition was to replace the rule of thumb approach to the allocation of work with something providing much greater levels of accuracy and certainty. He therefore developed a system of work study, a system of staff selection and a system of incentive payments all calculated to rationalize and standardize the production process (Shenhav 1999).

CONCLUDING REMARKS

The scientific revolution of the seventeenth century stimulated a great demystification of the natural world and its possibilities – a situation which would see science often in conflict with religion. The ideas of Locke, Newton and Hume provided an approach to science combining experimental method, the notion that our knowledge is always limited and the view that, in any case, what we construe as being knowledge from time to time is always tentative and perishable. This coincided with the rise of modernity – the process of industrialization, urbanization, standardization and rationalization which gradually spread around the world. The life and work of Josiah Wedgwood provides an example of the evolution of productive methods under the prompting of an empirical genius with an understanding of an emerging market. In the sphere of medicine, developments in anatomy provided an understanding of the human body which was truly revolutionary but also deeply disturbing – the idea of dissecting a human body being abhorrent to many people and running counter to religious sensibilities. Nevertheless, greater understanding of the human body stimulated changes in the approach to surgery and its possibilities. In this process we can detect the shift from basic or fundamental science to applied or practical science. A similar process occurred in engineering, placing an emphasis on technological development, and ultimately replacing the rule-of-thumb approach of the early, empirical engineers with the scientifically based knowledge of the university-trained professional engineers of the present day. This in turn prompted a new approach to the means of marshalling the available resources, which led to the emergence of modern management techniques.

CHAPTER 3

SCIENCE, R&D AND TECHNOLOGICAL DEVELOPMENT

INTRODUCTION

This chapter traces the linkages between scientific endeavour, applied science and technological progress through the mediation of commercially sponsored R&D, deploying insights derived from the philosophy of science. It also examines some of the organizational and management structures involved. Science is sometimes perceived as neutral and value free; dispassionate, rational and disengaged from contextual matters; concerned only with progress and the search for truth. However, even brief reflection will reveal that science is certainly not free of the context in which it is set and that scientists are no less creatures of their particular environment than practitioners in the social sciences and the humanities (Jardine 1999). Similarly, those aspects of scientific output which impact on the development of technology and the subsequent evolution of products and services are by no means neutral. Instead, they are subject to numerous external influences – not least the pursuit of strategic and economic advantage and the vagaries of the market. Before moving on to examine the ways in which applied science, R&D and technological development are linked, it is useful to consider the claims of two of the most influential philosophers of science: Karl Popper (1902–94) and Thomas Kuhn (1922–96).

THE PHILOSOPHY OF SCIENCE – THE POPPER/KUHN DEBATE

Karl Popper was responsible for the establishment of the philosophy of science as a distinct subject in its own right. He sought to distinguish between the claims of what he characterized as pseudo-science – such as Marxism and Freudianism – and 'hard' science as exemplified by Albert Einstein's (1879–1955) theory of general relativity. The key issue in Popper's view was what he termed falsifiability. According to Popper, Marxism and Freudianism share the same defect – neither set of theoretical propositions is capable of rigorous testing and possible refutation. Instead they are slippery – amenable to being adjusted to any set of circumstances in order to sustain the theory. In contrast, Einstein's theory is couched in terms which are patently amenable to rigorous testing. Popper had direct experience of

Marxism, having for a brief period been a Communist in post-First World War Vienna, before rejecting it in favour of his own brand of liberalism. His encounter with the ideas of Einstein was equally direct and formative. In his autobiography *Unended Quest*, published in 1974, he gives an account of attending as a young man a lecture given by Einstein in Vienna, being dazzled by Einstein's intellect and subsequently acknowledging him as the single greatest influence on his own work. Popper's two central texts are *The Logic of Scientific Discovery* (1934) and a collection of essays *Conjectures and Refutations: the Growth of Scientific Knowledge* (1963) but he also contributed to political theory with books such as *The Open Society and its Enemies* (1946). In defending the 'open society' against the contemporary incursions of totalitarianism, Popper argued for a liberal order in which, consistent with John Stuart Mill's book *On Liberty* (published in 1859), various ideas can contend in a sort of intellectual marketplace. In Popper's view scientists should produce theories which are readily amenable to refutation; science progresses through the constant process of error elimination as the frontiers of knowledge are pushed forward in the search for truth. He termed this approach critical rationalism. For Popper, the problem of induction which we noted in Chapter 2 becomes the key to scientific progress, constantly striving to go beyond our present understanding of the world by challenging the prevailing orthodoxy. He saw this competition between theories as literally a Darwinian struggle for survival in which inadequate theories are overthrown by theories of greater content (Okasha 2002). Whether Popper was successful in demonstrating that there is a clear demarcation between what he termed pseudo-science and hard science is problematical.

Popper faced a different challenge in the form of Thomas Kuhn's *The Structure of Scientific Revolutions*, the first edition of which appeared in 1962. Popper's work gave the strong impression that scientists are basically involved in a constant process of conjecture and refutation, searching for ever-greater levels of explanatory capacity in a sort of linear progression. Kuhn challenged this by suggesting that, for most of the time at least, scientists are engaged in what he described as 'normal science' conditioned by what he termed the prevailing paradigm. According to Kuhn a paradigm is a specific scientific outlook which is accepted by a scientific community at any specific time and in any specific place. For example, we have already encountered the theory of the humours in Chapter 2. Far from seeking to refute the prevailing paradigm, normal science tends to reinforce it, being largely concerned in what Kuhn described as puzzle-solving. Kuhn saw normal science as a highly conservative activity with its practitioners content to work within the prevailing paradigm rather than striving to refute it. According to Kuhn, radical change only happens when numerous anomalies occur which cannot be explained by the prevailing paradigm. Eventually confidence in the paradigm is exhausted and normal science grinds to a halt. At this point alternatives to the old paradigm are proposed, one of which will ultimately prevail, and the scientific community will eventually convert to the new paradigm, thereby completing a scientific revolution. Although there is ultimately no way to determine which view is correct, the Popper/Kuhn debate does shed light on the nature of scientific endeavour. Finally, as will be seen in the next section, there are certain similarities between what Popper proposes and the nature of basic research and Kuhn's claims and the nature of applied research.

FROM APPLIED SCIENCE TO R&D AND TECHNOLOGICAL PROGRESS

It may be that there are individual scientists working in solitary isolation from the world. However, the bulk of science and perhaps all R&D are carried out in an institutional setting – whether that setting is academic or industrial. Kressel (2007) has usefully identified three distinct kinds of R&D and offers the following definitions:

- *Basic research* – which seeks to discover the laws governing natural phenomena and is most likely to be successful when carried out in a free and open environment untrammelled by social or organizational constraints. In this situation researchers can follow their instinctive or intuitive hunches with the possibility that significant discoveries may emerge. This is a high-risk/high-cost approach but may yield enormous gains if successful. Kressel uses as an example the initial research in the 1930s and 40s on semiconductors which produced immense technological and commercial results.
- *Applied research* – which takes place as part of the process leading to 'productization'. Applied research starts with knowledge of basic physical phenomena and seeks to identify a route to their practical utilization. Kressel uses the example of the work carried out at Bell Labs which made it possible to prepare single-crystal germanium and establish its properties. This, in turn, led to the fabrication of the first transistor.
- *Product development* – which covers all the various activities leading to the creation of new products or processes, help enhance existing products, or improve their 'manufacturability'. Kressel uses the example of the development work which was carried out to find the appropriate methods to effectively fabricate transistors.
- In Kressel's opinion, and he is surely correct, innovation is not possible without basic research, which leads us to the subject of research and development – R&D – commencing with the contribution of the inventor and entrepreneur Thomas Edison.

INDUSTRIAL R&D – THE EDISON, BELL AND RCA LABORATORIES

Thomas Edison (1847–1931) was the self-taught inventor and developer of numerous devices (he held over 1,000 US patents) in the spheres of electrical engineering and mass communications. He is also credited with establishing one of the world's first industrial research laboratories where he recruited a research team with the necessary abilities to develop and refine his numerous inventions. Finally, he was able to make his inventions available to a mass market by deploying the principles of mass production pioneered by his friend, and sometime employee, Henry Ford. Edison succeeded in bridging the cultural gap between so-called pure science and the practical inventors – technologists who did applied research and product development. In 1876 he opened his laboratory in Menlo Park, New Jersey, which he called the Invention Factory. There scientists and technologists were integrated into a single organization. In 1887 he moved to new premises – the Edison Laboratory – in West Orange, New Jersey, creating the world's first R&D centre where breakthroughs were made in many areas including the production of the phonograph, motion picture technology, incandescent lighting, and electric power generation.

Edison's central concern was with commercial success. R&D for its own sake did not interest him and the focus was firmly on the development of product innovation. The combination of teamwork and the creation of successful products made Edison's laboratory the forerunner of all subsequent research strategies and inspired the formation of several large corporate laboratories in the early twentieth century, including those at American Telephone and Telegraph (AT&T), International Business Machines (IBM), the Radio Corporation of America (RCA), Westinghouse, and General Electric. The largest of these was AT&T's Bell Laboratories, founded in 1924 (Kressel 2007). Bell Laboratories (Bell Labs) was officially founded by the physicist Frank B. Jewett (1879–1949) in New York in 1925, although its origins stretch back to 1912 when Jewett assembled a team of physicists and engineers to work on problems associated with long-distance telephony (Shurkin 2006). The research undertaken at the Bell Labs spanned every aspect of R&D from basic research through to product development. In 1947 William Shockley (1910–89), John Bardeen (1908–91) and Walter Brattain (1902–87) jointly invented the transistor at Bell Labs. This was possibly the single most important event in twentieth-century electronics, making feasible the subsequent development of the integrated circuit and the microprocessor, and its significance was duly recognized by the joint award of the Nobel Prize for Physics in 1956 (Shurkin 2006). Bell Labs operated on a huge scale under the protected arrangements AT&T had with the US government. By the time AT&T was broken up in the 1980s, through agreement between the company and the US Department of Justice, Bell Labs had some 20,000 scientists, engineers and support staff distributed around the USA. AT&T was organized on the basis of vertical integration with the company having control of the US telecommunications network from technology to the production of equipment and, via the Bell operating companies, the bulk of service delivery. AT&T's profits were guaranteed by cost plus pricing negotiated with government regulators and based on assets deployed. This situation provided an environment in which long-term research projects could thrive and research results could be turned into products. Bell Labs pioneered modern wire-line and wireless telephony systems and established the scientific and technical foundations for the semiconductor industry (Gertner 2012). The Labs also made major contributions to computer science, including the development by Dennis Ritchie (1941–2011) and his colleagues of Unix, arguably the most important system for large-scale and high-performance computing. Unix has been aptly described as 'the best screwdriver ever built' and provides the operating system for many of today's major corporate data centres such as Amazon and Google. Steve Jobs (1955–2011) used Unix as the basis for his NeXT computer workstation, and it later became the foundation for Apple's smartphones and other products. The Unix philosophy of free access inspired the so-called open source software movement and its Unix variant, Linux, which now powers most of the servers on which the Internet depends (Gertner 2012).

Strong corporate leadership and vision are essential elements in generating successful innovation research. A key example is RCA Laboratories under the leadership of David Sarnoff (1891–1971). Sarnoff, unlike Edison, was a businessman rather than a technologist and it was under his direction that the pioneering work on colour television was carried out. In addition to its research and manufacturing capacity, RCA also owned a broadcasting network, the National Broadcasting Company (NBC) via which it could launch a colour television service – it did so in 1951. RCA Laboratories invented and patented every one of the core technologies, including manufacturing methods, for colour television. The technology was designed to be compatible with existing TVs and RCA adopted a lucrative licensing strategy which permitted other manufacturers and broadcasters to enter the market. RCA made huge financial investment to achieve this situation and it was ten years before the company began to earn profit from its colour television receivers (Bilby 1986). It was profit from its television products which enabled RCA to expand into many other product areas – for example, the first germanium transistor designed for consumer electronics was developed at the RCA labs in 1952 and manufactured at the newly established Solid State Division in 1953. RCA were also active in the recording industry, manufacturing gramophones and developing recording technology. In 1949 the company issued the first seven-inch, 45 rpm record and during the 1950s they became leading players in the record industry when they signed Elvis Presley (1935–77) from Sun Records of Memphis, Tennessee for the then massive sum of US$35,000. In spite of serious misgivings among some RCA executives, Presley's first recording for the company, *Heartbreak Hotel*, was released in 1956 and became a number one hit. By the end of a six-week period it had sold eight million records and by

1957 Presley had become an annual US$20 million industry – big business indeed! (Cohn 1969). *Elvis Presley*, the album which followed *Heartbreak Hotel*, was largely recorded in RCA's New York studio. It was a massive success, becoming the first rock-and-roll record to reach the top of the *Billboard* chart and remaining there for ten weeks. According to the Recording Industry Association of America, Elvis Presley has sold 2.5 billion records. Self-evidently product, in terms of entertainers and entertainment, is as important as the hardware. In fact the two work together, feeding off each other, and facilitating the rise of new markets and profitable possibilities. In the case of Elvis Presley it was a combination of raw talent and sophisticated technology at a price kids could afford.

HENRY FORD AND MASS PRODUCTION

Henry Ford was born in 1863 in Dearborn, Michigan where his parents were farmers. From his earliest years he was fascinated by all kinds of machinery. Ford left school at the age of 17 and, ignoring his father's wish for him to become a farmer, became an apprentice engineer. When he completed his apprenticeship, he joined the local representative of the Westinghouse Company working on the construction and maintenance of steamdriven road engines. Ford soon decided that steam was not suitable for lightweight vehicles and turned his attention to the internal combustion engine. During the years of his apprenticeship Ford had read about the silent gas engine in a copy of a British journal, *World of Science*. This was probably an account of the engine invented by Nikolaus August Otto (1832–91) and, in 1885, Ford got the opportunity to examine such an engine when he repaired one at the Eagle Iron Works in Detroit. In 1887 he constructed a fourstroke engine based on Otto's design (Sheldrake 2003). He also built a workshop and continued his experimental work on the internal combustion engine in his spare time, constructing a doublecylinder engine in 1890. Ford worked at the Detroit Electric Company as an engineer and machinist and later joined the Edison Illuminating Company where he rose rapidly to become chief engineer. Between 1893 and 1895 he constructed his first automobile which, although the ignition and transmission were primitive, was nevertheless capable of running at two speeds – either 10 or 20 miles an hour (Brinkley 2004). Ford left Edison Illuminating in 1899, becoming chief engineer and a minority stock holder of the Detroit Automobile Company. The venture was a failure and, although the firm survived to become the Cadillac company, Ford quit and established the Ford Motor Company in 1903.

Between the foundation of his company and the emergence of the Model T in 1908, Ford produced a variety of automobiles. The popularity of the Model N, at a selling price of US$600, enabled Ford to rationalize production and finance the development of a successor car – the Model T. The Model T embodied every feature required of the all-purpose, volume automobile which Ford had long sought to produce. Mechanically simple and easy to service, it could cope with the rough road and off-road conditions encountered in many areas of the USA. The Model T was an immediate success and, having created the automobile, Ford began improving manufacturing techniques so that greater numbers could be produced and unit costs lowered. He situated as much manufacturing capacity as possible on a single site, reducing transport costs and also dependence on outside suppliers. He hired talented individuals and deployed their gifts on the problems of meeting what rapidly became an insatiable demand for Model Ts. Among those employed by Ford were Charles E. Sorensen (1888–1961) and Clarence Avery (1882–1949). By 1909 these men were collaborating with Ford himself on the development of the continuous flow of materials through Ford's Piquette Avenue plant in Detroit. It was there that the first experiments were made with the famous moving assembly track, beginning with magnetos and later extending to other parts of the Model T. In 1910 the Ford Motor Company moved into its new Highland Park plant designed by the prominent industrial architect Albert Kahn (1869–1942). In 1913 full-scale assembly line production of Model Ts began at Highland Park with astounding results – output rose from 8,000 vehicles in 1907 to a quarter of a million in 1914; profits from the US$1 million that the Model N earned to US$27 million for the first year of the mass produced Model T. Unit costs reduced to the extent that, by 1916, Ford touring cars could be sold for less than US$400. This achievement has been characterized as one of the great forward leaps in the history of technology. The techniques of mass production became known in Germany as Fordismus and Aldous Huxley (1894–1963) even cast Ford in the role of deity in his 1931 novel *Brave New World* (Rae 1959).

The three core elements in the Ford system were accuracy (including the standardization and interchangeability of parts), continuity (the moving assembly track) and speed (the careful correlation of manufacture, material handling, and assembly) all resulting in the achievement of flow. A modern example of this combination can be seen in containerization in shipping examined in Chapter 6. The production system which evolved at Highland Park developed in a piecemeal fashion, commencing with the installation of conveyor belts and gravity slides for materials being transported to the machine shop. Meanwhile, in the foundry, an endlesschain carrier was erected which passed the moulds under spouts that filled them with molten metal. Continuous conveyor belts were installed to transport components to the assembly lines. The success of the Model T provided Ford with an immense inflow of cash and a technological and commercial edge over his rivals which, for the while at least, seemed unassailable. In 1923 the company held over 50 per cent of the American automobile market with sales of 1.7 million vehicles compared to General Motors (GM) with 800,000. By 1927, however, it had been overtaken by GM, making heavy losses and burdened with a product which no longer met the aspirations of increasingly discerning consumers. On 26 May 1927 the last of some 15,000,000 Model T's rolled off the assembly track and the biggest selling car in history was finished. Not until 1928 did the company return to the market with the Model A which partially restored its fortunes. However, the Ford

Motor Company no longer led the field in automobile design and production, that role having been usurped by GM under the guidance of its innovative president and chief executive, Alfred Sloan (1875–1966).

ALFRED SLOAN AND THE SLOAN STRUCTURE

Sloan was born in 1875 in New Haven, Connecticut. He graduated from the Massachusetts Institute of Technology in 1895 with a degree in electrical engineering and took a job as a draughtsman at the Hyatt Roller Bearing Company in Harrison, New Jersey. When Hyatt got into financial difficulties Sloan's father and a group of associates raised sufficient funds to keep the company going and Alfred took a leading role in restoring the company's fortunes (Sheldrake 2003). In 1916 the founder of GM, William C. Durant (1861–1947), proposed to Sloan that Hyatt be sold to the United Motors Corporation, a holding company which Durant had set up with the purpose of acquiring leading components makers with a view to achieving greater vertical integration and ensuring a steadier flow of supplies. Sloan's initial reaction to the proposal was one of scepticism. Hyatt's customers included the Ford Motor Company as well as GM and, under Sloan's guidance, the company had achieved the combination of large-scale production and accuracy necessary to meet the demands of the burgeoning American automobile industry. However, on reflection, Sloan realized that Hyatt had paradoxically become too big to be independent (Rae 1959). Having geared itself up for mass production it could only operate profitably in a stable market dominated by large corporations such as Ford and GM. If either of these companies chose to manufacture their own bearings then Hyatt would be left in a difficult situation. Thus Durant's proposal was accepted and Sloan became the president of United Motors and a major stockholder in the company.

When United Motors was absorbed into GM in 1918, Sloan combined the jobs of vicepresident of GM with that of president of United Motors. GM was created by Durant from many automobile and component producers but he was incapable of imposing an orderly structure on the enterprise, a weakness cruelly exposed by the economic crisis of 1920. At the end of the First World War, the US economy enjoyed a brief boom which stimulated a general mood of optimism among automobile manufacturers. Demand and prices rose quickly and new automobiles were in short supply. However, during the spring of 1920 demand began to fall and continued to decline until manufacturers, starting with Ford, were forced to cut prices. Although the depression was brief the crash of 1920 had a lasting impact on the automobile industry and generated major rationalization and restructuring (Rae 1959). At GM the fall in demand soon caused serious cashflow problems. In spite of Sloan's urgent pleas to cut prices, Durant refused and by October 1920 the company's managers were struggling to find enough cash to meet immediate needs such as invoices and payrolls. Meanwhile the value of the company's stock collapsed and Durant made a reckless attempt to sustain its value by buying GM stock on credit. The initiative failed, forcing Durant to resign as president of the company he had founded. GM was only saved from extinction by a huge infusion of cash from the Du Pont Corporation which held a major financial stake in the automobile company. Pierre du Pont (1870–1954) came out of semiretirement to become president of GM, one of his first executive acts being to name Alfred Sloan as his successor.

When Sloan joined United Motors in 1916 he began devising an appropriate organizational structure. As each of the constituent companies was in capable hands, Sloan was able to concentrate his energies on establishing a general office charged with coordinating and expanding the operations of each company as appropriate. Among the innovations Sloan introduced were uniform accounting procedures and a marketing organization. He also worked to ensure that GM's dealers received a reliable supply of spare parts and accessories. However, although Sloan was successful in organizing United Motors, he became increasingly concerned about GM's lack of structure and system. He had no idea, for example, how United Motors fitted into GM at large. Further, he was not certain how the various operating divisions of GM itself supplemented or complemented each other. At the end of 1919 Sloan began working on an *Organization Study* which embodied the notion of decentralization and effective delegation of authority which eventually became basic to his philosophy of industrial management (Rae 1959). At the beginning of 1920 he submitted his plans for the reorganization of GM to Durant, who approved them in principle but showed no appetite for arranging their implementation. With the arrival of Pierre du Pont, however, a wholly fresh approach was adopted and Sloan's study formed the foundation of GM's management policy (Sloan 1963). The core of Sloan's new structure for GM was the principle of decentralization. In one sense this was easy to achieve, given the lack of company-wide cohesion during the Durant years. Sloan's notion of organization therefore involved the development of co-ordination without sacrificing whatever benefits accrued from the existing decentralization. Basically, Sloan established the staff-line pattern of a military organization. The separate companies were transformed into operating divisions, each under an executive with virtually complete responsibility for its management. The staff functions, such as R&D, financial management and marketing were organized separately, with their services available to the whole GM family but without direct authority over the operating divisions. Sloan made it possible for GM to be transformed from a loose conglomeration of automotive and other enterprises into a coherent and co-ordinated business structure, combining to a unique degree the advantages of concentration with the flexibility of decentralization (Rae 1959).

The new structure was put in place at GM during the years 1921 to 1925 – one of the earliest tasks undertaken by Sloan and his executives being to define boundaries for the activities of each division. As well as failing to produce a coherent organizational structure, Durant had not really attempted to create a rational product range and the various divisions of GM competed against each other rather than their external competitors. Under Sloan's direction the product range and market orientation became defined – Cadillac sold in the highest price bracket with Buick next, followed by Oakland and then Olds, with Chevrolet in the largest-volume, lowest-price market. In 1925 GM introduced the six-cylinder Pontiac to fill the gap in its range between the Oldsmobile and the Chevrolet and also purchased the Yellow Cab Manufacturing Company of Chicago, merging it with its own truck division to form the Yellow Truck and Coach Manufacturing Company. Finally, GM began to acquire foreign subsidiaries including Britain's Vauxhall Motors in 1925 and Germany's Adam Opel Company in 1929. The decentralized organizational structure which Alfred Sloan devised for GM was hugely influential and was subsequently emulated throughout the industrialized world. Decentralized operations with centrally coordinated control enabled the separate elements of GM to work effectively towards a common goal. Sloan's innovations in the sphere of organizational structure effectively complemented those of Henry Ford in the sphere of manufacturing (Sheldrake 2003).

THE FURTHER DEVELOPMENT OF MODERN MANAGEMENT

Early industrialization was dominated by owner-managers, entrepreneurs who took responsibility for the entire range of activities undertaken by the enterprise – indeed Henry Ford continued with this approach, with dwindling success, deep into the twentieth century. Increasingly, however, the scale of operation and complexity of business organizations determined that professional managers, rather than owners, became responsible for running corporations. Owners therefore delegated the running of the business to managers – a prime example, as we have seen, being Alfred Sloan. Probably the first attempt to provide a systematic account of what managers do was made by the French industrialist and mining engineer Henri Fayol (1841–1925) in his 1916 book *General and Industrial Management*. He identified the five elements of management as *planning* – attempting to assess the future and making provision for it; *organizing* – ensuring that the organization is provided with everything necessary to its functioning, including raw materials, equipment, capital and staff; *command* – obtaining the optimum return from the staff in the interests of the whole concern; *co-ordination* – ensuring the harmonization of all the elements involved in the organization such as sales to production, expenditure to income, and stocks to consumption and, finally, *control* – ensuring that everything which occurs in the organization conforms with the adopted plan. Whereas Taylor theorized from the base of the organization, at the level of the individual worker, Fayol's position was far more strategic – operating from the top down.

Although complex organizations existed in earlier, and even ancient, times it was the coming of the railways in the nineteenth century which prompted the establishment of recognizably modern organizational structures. As has been noted, industrial concentration (particularly in engineering and transport) and growth in the scale of operation gradually eroded the centrality of the individual entrepreneur/manager. Even if the individual continued to lead the organization, more and more work was delegated to professional managers (Galbraith 1967). Such managers derived their authority not from ownership or family connection but from a combination of expertise and status in the organizational hierarchy. By the time of the First World War industrial countries such as Britain, France, Germany, Japan and the USA already possessed large complex organizations. In Germany the pace of industrialization and industrial concentration was particularly rapid in the years following unification in the 1870s. It was there that the sociologist Max Weber (1864–1920) developed his theory of bureaucratic organizational structure, which has remained influential to the present day. Bureaucracy is associated with large-scale operation and can be defined as a form of organization characterized by specialization of labour, a specific authority hierarchy, a formal set of rules, and rigid promotion and selection criteria. Weber believed that in modern organizations legitimacy most often derives from *legal authority* and a system of abstract rules to which all members of the organization are subject. In a bureaucracy, authority is not inherent to the person holding it but derives from the impersonal order that appointed him or her to the job. Work should be organized on a continuous, regulated basis and divided into distinct professional spheres. Jobs will be arranged hierarchically with a clear line of reporting or command (on a quasi-military pattern). Finally, administration will be based on written documents embodying agreed procedures, thus tending to make the office (or bureau) the hub of the modern organization. In theory at least, bureaucracy should provide the advantage of certainty of performance, rationality, transparency, fairness and consistency. In simple terms, bureaucracy is the administrative equivalent of scientific management, placing maximum emphasis on control and allowing minimal scope for individual autonomy.

As we have seen, Henry Ford adapted the techniques of scientific management to the mass production of automobiles. Ford concentrated as much of the necessary manufacturing capacity as possible on a single site and thereby reduced transport costs and limited dependence on outside suppliers. To this he added the crucial moving assembly track to facilitate the continuous flow of materials through the plant. All of this came together in the manufacture of the world's first mass-produced car, the Model T. Ford plants were cloned in Europe and his pioneering approach to manufacturing emulated throughout the industrialized world. One outcome was that automobiles ceased to be a luxury item for the wealthy only and became increasingly available to all consumers. The Model T remained in production for some 20 years, becoming the biggest-selling car in history, challenged only by the Volkswagen Beetle – a vehicle inspired by the Model T and Ford's production methods. Ford's methods remain influential and are deployed not merely in the manufacture of automobiles but in many spheres of mass production, including fast food outlets such as McDonald's and the manufacture of the simple (only eight moving parts) AK47 assault rifle – perhaps 80 million weapons currently in circulation. In every case the system is built on close managerial control of the production process. Similarly the organizational structure devised by Alfred Sloan at GM was hugely influential and widely adopted throughout the industrial world.

CONCLUDING REMARKS

By the early twentieth century scientific research was being systematically harnessed by industry. In the developed nations, and most prominently the USA, a clear linkage had developed between entrepreneurial activity, innovation and productive effort within a market-based economy which, whatever its inherent faults, has been remarkably successful. Increasing scale of operation generated organizations displaying the combined techniques of scientific management, mass production and bureaucracy. The management thinker Henry Mintzberg claimed (1989) that the defining characteristic of such organizations is standardization. Large operating units, extensive division of labour, and above all, the need to maintain management control, all tend to what Mintzberg describes as Machine Bureaucracy characterized by a combination of detailed operating procedures, routine patterns of operation, highly developed sets of rules and procedures and formal communication channels, centralized decision making and a clear distinction between line and staff workers. Mintzberg did not take machine bureaucracy to be the universal configuration for all industrial and business organizations. Instead he depicted it as being of limited application, appropriate only in those organizations where the work is routine, repetitive and relatively simple in nature. Among the kinds of organization Mintzberg identified as suitable candidates for machine bureaucracy were a national post office, a custodial prison, an airline, a steel company and a giant automobile company. He could have added central and local government and large financial institutions. In contrast to the standardized activity associated with machine bureaucracy, Mintzberg identified an organizational configuration most appropriate to innovation which he described as adhocracy.

Innovation demands a degree of flexibility and creative space which is impossible to achieve within the structural confines of scientific management, mass production and bureaucracy. Innovation is often orientated around small project teams of highly qualified professionals working semi-autonomously with minimal central direction (Mintzberg 1989). This is the natural habitat of the project manager, operating across conventional professional boundaries – influencing rather than commanding and sustaining high levels of communication – a world of colleagues rather than bosses and operatives. Major corporations will often possess both organizational configurations – machine bureaucracy and adhocracy – within the overall structure. Combining innovative energy with organizational giantism is a key issue for major corporations. With the demise of the Model T, the Ford Motor Company surrendered its lead in the automobile industry to GM and, as will be seen in the following chapters, there are numerous examples of businesses which, having fought their way to market leadership, have been replaced by newcomers once their innovative edge has been dissipated.

CHAPTER 4

INNOVATORS AND ENTREPRENEURS

INTRODUCTION

This chapter examines the nature of entrepreneurship and innovation. We live in a world of huge organizations. Civil aviation is dominated by two major producers and automotive by four or five, as are petrochemicals, pharmaceuticals, financial services and consumer electronics. Wal-Mart stores, on many measures the world's largest company, employs 1.5 million people, while HSBC employs 200,000 and operates in 80 countries. McDonald's have over 30,000 restaurants in 118 countries and serve 64 million people every day. A few major corporations (such as Microsoft, Intel, Apple and Samsung) are pre-eminent in the world of information technology, while Amazon, Facebook and Google have become household names. In line with all of this we have become comfortable with a marketplace dominated by brands and constantly stimulated by advertising. Companies compete with each other to become and remain market leaders and consumers display brand loyalty or not depending on their whim. At the base of all this activity will be an original innovation, whether it be a hamburger or a jet engine, and often a single entrepreneur with the vision and drive to become and remain a market leader. Later in the chapter we examine the examples of McDonald's, Intel and Microsoft.

Innovation can be usefully defined as an idea or an object or a practice that is perceived as being novel by an individual, a group or a society (Rogers 2003). Kressel views innovation as an irreversible shift in the way we do things, involving not just products but also techniques, forms of organization and the location of new markets. He identifies two forms of innovation which he defines as revolutionary innovations (which generate far-reaching and often unforeseen consequences) and evolutionary innovations which translate these radical concepts into marketable products and services (Kressel 2007). Finally, the management thinker Peter Drucker (1909–2005) in his book *Innovation and Entrepreneurship: Practice and Principles*, first published in 1985, made a clear link between innovation and the role of entrepreneurs who perceive change as the opportunity to supply and profitably exploit a new product or service. Both Kressel and Drucker were strongly influenced by the Austrian economist Joseph Schumpeter (1883–1950) and the process of what he termed creative destruction.

CREATIVE DESTRUCTION – THE DRIVING FORCE BEHIND CAPITALISM

In a series of books, culminating in his *Capitalism, Socialism and Democracy* first published in 1942, Joseph Schumpeter sought to discover or elucidate the underlying dynamics of capitalism and its future trajectory. In Schumpeter's view capitalism is characterized by relentless change – intrinsically dynamic in nature, ever shifting and churning as old processes and enterprises decline and newcomers take their place. This is the process of creative destruction which Schumpeter claimed is an essential characteristic of capitalism – basically perpetual dynamic change serving to revolutionize the economic structure from within. New technologies, new products and modes of transport, new markets and novel forms of organizational structure emerge, incessantly fracturing and sweeping away the existing order. Although attempts might be made to inhibit this process, ranging from state intervention to trade union organization, the outcome is ultimately the same. Schumpeter also sought to identify the source of profit within the capitalist enterprise. In his book *The Theory of Economic Development*, first published in German in 1911 and in English in revised form in 1934, he allocated this role not, as might be expected, to the exploitation of labour or the return on capital employed, but to the activities of entrepreneurs.

According to Schumpeter, successful entrepreneurs must possess the necessary combination of vision and leadership to identify the trajectory of change and, so to speak, ride it. Whereas normal business will follow established routines, in contrast an entrepreneur is an innovator with the ability to combine the factors of production in novel ways (Heilbroner 2000). Schumpeter thus saw entrepreneurs as a sort of elite struggling against ever-increasing intervention by the state and other agencies which only serve to inhibit the entrepreneurial impulse, thereby slowing the pace of innovation – and, by extension, human progress. On this view, innovation ultimately becomes tamed, institutionalized and reduced to mere routine; bureaucratic structures are erected which make innovation ever less likely; while entrenched, vested interests inhibit the pace of change – issues highlighted by Rosabeth Moss Kanter in her 1983 book *The Change Masters*. However, whereas Schumpeter was interested in articulating the characteristics of an ideal type, Kanter was concerned with what she claimed was the USA's alarming loss of innovative energy caused, as she saw it, by a combination of complacency and a startling decline in the entrepreneurial spirit. According to Kanter, the remedy was to be found in what she termed idea power, whereby traditional, bureaucratic low-trust organizations might be transformed into modern, high-trust organizations with a total commitment to innovation (Sheldrake 2003). These are issues which will be explored further in the chapters which follow. Meanwhile we turn to the topic of innovation management.

THE ROLE OF INNOVATION MANAGEMENT

It was claimed in Chapter 3 that strong corporate leadership is an essential element in generating successful innovation research. A similar claim can be made for the successful exploitation of a specific innovation and the bringing of a novel product or service to the market. According to Peter Drucker successful innovation must always aim at leadership in the chosen sphere of activity. This requires the development of an appropriate strategy – whether it is calculated to establish dominance in a particular industry or merely identify and exploit a niche in the market. In Drucker's view any entrepreneurial strategy must attain leadership or it will merely create an opportunity for competitors (Drucker 1985). Opinion is divided as to whether large companies can remain sufficiently nimble to nurture continuous innovation. Certainly, large companies can deploy greater resources if they stay hungry and resist the process of innovation inhibiting bureaucratization which Wu has aptly described, in the context of the information technologies, as 'the Cycle' (Wu 2010). On this view it is often possible to identify the origins of an innovation as more or less an individual's hobby or private passion. Wu cites the example of the inventor Alexander Graham Bell (1847–1922) who is credited with the development of the first practical telephone. From tiny beginnings in the 1870s, by the 1900s Bell's initial invention had generated a massive industry, organized on the basis of vertical integration under the leadership of Theodore Vail (1845–1920) an advocate of the 'closed system', basically monopoly power, as the means to supply a universal service.

This evolution from simple beginnings to complex mega-organization can be identified in many industries and runs the risk of replacing the original creative spark with routine activity dominated by habit and corporate vested interests. An organization needs therefore to be constantly redefining itself; asking itself questions about which market/s it is in and treading a fine line between diversification and sticking with what it does best. In the long run most business organizations fail because the 'swarm' of competition catches up or new innovations become dominant, eclipsing the demand for an existing product or service. Equally, it is contentious whether a monopoly position supports and stimulates innovation by providing the necessary resources to sustain meaningful R&D or stifles it by giving undue shelter from healthy competition. Further, the role of the state in nurturing innovation remains a matter of debate – whether the state has a positive effect or a negative one – whether the state should let entrepreneurs make the running or become involved in trying to pick winners.

It is often claimed that the most innovative ideas are generated by start-ups – there are numerous examples of small lean groups turning an idea into product and then either taking it to the market or selling it on to a bigger company for development. However, it is also the case that success is elusive, that most new products do not succeed and that most start-ups fail. In his 2011 book *The Lean Start-up*, the successful Silicon Valley entrepreneur Eric Ries sets out what he terms the five principles of lean start-up as follows:

- *Entrepreneurs are everywhere* – a start-up is not necessarily a new venture set up in someone's garage. It can be, for example, a unit in an existing organization, specifically established to create new products and services. According to Ries entrepreneurs can be found in any organization, including large corporations, in any sector or industry.
- *Entrepreneurship is management* – a start-up is an organization, not merely a particular product, and requires a specific style of management geared to conditions of extreme uncertainty. Self-evidently this will not be bureaucratized – control needs to be matched with high trust and we have already identified Mintzberg's adhocracy as the appropriate organizational configuration for innovative activity. Nathan Myhrvold, former Chief Technical Officer of Microsoft, once described this approach in my hearing as 'no police'.
- *Validated learning* – start-ups exist not just to make products or generate cash or even serve customers. They also have to learn how to build a sustainable business which requires exposing the venture to regular monitoring or reality checks to test the validity of the original vision.
- *Build–Measure–Learn* – start-ups need to create a rapid process of feedback in order to establish whether results match the original vision. Feedback provides the essential information required to proceed with an idea or to change tack.
- *Innovation accounting* – conventional management systems need to be put in place, including financial accounting (what Ries calls the boring stuff) in order to improve entrepreneurial outcomes and hold innovators accountable. In a sense this is just straightforward project management – applicable to tiny start-ups and huge industrial enterprises alike.

Basically, the process of innovation needs to be managed in such a way as to nurture individual initiative while meeting corporate product requirements, whether the corporation be a tiny start-up or a massive transnational. Henry Kressel, for many years a researcher and manager at the RCA Labs where he was responsible among other things for the commercialization of products, took the view that the innovators should be as closely involved as possible with the management of innovations into viable products, including the process of ensuring manufacturability (Kressel 2007). This is surely sound advice and illustrates the obvious value of team working and the development of mutual understanding across professional boundaries.

RAY KROC AND MCDONALD'S

In his book *Understanding Organizations* the management thinker Charles Handy examines the topic of organizational culture – the values and traditions which turn organizations into what he describes as cohesive tribes (Handy 1999). He also notes that organizations most often bear the image of their founder and are sustained by a foundation myth which articulates the heroic story of a company's origins. This is certainly true of the McDonald's Corporation and its foundation by the entrepreneur Ray Kroc (1902–84). The original McDonald's was opened by Patrick J. McDonald in 1937 on Huntington Drive (Route 66) near the old Monrovia Airport in Los Angeles County, California. Called the 'The Airdrome' it sold hamburgers for ten cents and all-you-can-drink orange juice for five. In 1940, Patrick McDonald and his two sons Maurice and Richard (known as Mac and Dick) physically relocated the building 40 miles east to San Bernadino. They renamed the restaurant 'McDonald's Bar-B-Q' and began serving a menu of twenty five different barbecued dishes. Eventually the McDonald brothers realized that the bulk of their profits came from selling hamburgers and decided to streamline their menu, offering a much reduced choice consisting of just hamburgers, cheeseburgers, French fries, apple pie, soft drinks and shakes. At the same time they reorganized their kitchen on an assembly line basis and moved to a self-service operation which they called the 'Speedee Service System'. In December 1948 they again changed the name of the restaurant, this time to plain 'McDonald's' and, based on its success, began franchising their idea – starting in 1953 in Phoenix, Arizona and Downey, California. The latter remains the oldest McDonald's restaurant still in operation, having been acquired by the McDonald's Corporation in 1990 and refurbished in original 1950s styling.

In 1954, Ray Kroc, then working selling Multimixer milkshake machines, learnt that the McDonald brothers were using eight of his machines (each of which could make five milkshakes at once) in their San Bernardino restaurant. Intrigued as to why a small hamburger joint was selling so many shakes, Kroc drove to the McDonald's restaurant to take a look at what the brothers were doing (Schlosser and Wilson 2006). Impressed by what he saw he was convinced that the brothers had discovered a winning formula capable of being franchised across the entire United States. Reluctant to shoulder an additional burden, the McDonald's brothers agreed that Kroc should undertake the task and he duly opened his initial McDonald's restaurant in Des Plaines, Illinois near Chicago in April 1955. On the same day he incorporated his company as McDonald's Systems, Inc. – later renamed the McDonald's Corporation. Early progress was slow but, by 1959, the company's spectacular growth had started, there were McDonald's restaurants in over 100 locations and Ray Kroc had begun to prefect his business model. He maintained a delicate balancing act between the imposition of tight, organization-wide standards and the fostering of an entrepreneurial spirit which encouraged ideas from all levels. Kroc deployed the concepts and techniques of scientific management and mass production to the restaurant business – the delivery of fresh, fast food (Gross 1996). Franchising was central to Kroc's success in building McDonald's into a major international brand, enabling the franchisees to buy into the business model of the parent corporation. The franchisees operate under the lead company's brand, taking their product from the lead company, enjoying the advantages of economies of scale and benefiting from the lead company's advertising campaigns – which in McDonald's case are substantial. McDonald's currently employs around 1.7 million staff (it is claimed that as many as one in eight Americans have at some time in their life worked for the company) and has annual sales estimated at around US$30 billion – 80 per cent of the restaurants are operated as franchises. The initial vehicle for Kroc's success was not the restaurants themselves but a subsidiary, the Franchise Realty Corporation, which enabled him to buy likely sites for future restaurants and then lease the sites to franchisees. This not only provided an alternative source of revenue, but also enabled him to impose a standardized corporate image – ensuring that the design of the restaurants, the menu and the service were identical throughout the network.

In 1961 Kroc bought the McDonald's brothers out of the business for US$2.7 million and began the process of professionalizing his expanding operation. A training programme was established which eventually evolved into the Hamburger University at Elk Grove, Illinois where franchisees and other staff were trained in the appropriate, scientific methods for operating a successful McDonald's outlet. Kroc also established an R&D facility dedicated to technical improvements in the processes of freezing, preparing, cooking and storing of food and the creation of new products. Although by the time Kroc died in 1984 he had built McDonald's to become the largest restaurant company in the world, he did not of course achieve this alone. Specifically he was assisted by perhaps the ultimate company man – Fred L. Turner (1933–2013) sometimes called 'the man who made McDonald's'. Kroc believed in promoting men from the shop floor and Turner began his career with the company on leaving the US Army in 1956 operating the grill in Kroc's original restaurant in Des Plaines. He rose rapidly through the ranks, becoming operations manager in 1958, president and chief administrative officer in 1968, taking over from Kroc as the company's chief executive in 1974 and remaining active at the top of the organization until his final retirement in 2004. Turner presided over the rapid expansion of the company, introducing new products and moving into overseas markets including the UK in October 1974. Turner was particularly active in the establishment of the first McDonald's Hamburger University, mentioned above, which was renamed the Fred L. Turner Training Centre in 2004. The McDonald's model has been adapted over the years to incorporate greater regional variations in the contents of the menu and the design of the actual restaurants. However, the core elements of Kroc's system remain in place based on the company's gospel of Quality, Service, Cleanliness and Value and, in spite of sustained criticism, most potently by Eric Schlosser in his 2004 book *Fast Food Nation*, the company continues to prosper.

GORDON MOORE, ROBERT NOYCE AND INTEL

We noted in Chapter 3 that Bell Labs were for many years the world leaders in industrial R&D and also that it was a team of researchers gathered there around William Shockley which did the pioneering work on the transistor; work for which Shockley, together with his associates John Bardeen and Walter Brattain, received the Nobel Prize for Physics in 1956. Even in Shockley's moment of triumph, however, there was a certain sense of disappointment in that the primary breakthrough was achieved by Bardeen and Brattain rather than himself (Shurkin 2006). Also, Shockley felt slighted when Bell promoted Bardeen and Brattain's names ahead of his own in patenting the transistor, became increasingly disillusioned with corporate life and began to plan ways to break away and run his own show. He took a sabbatical from Bell Labs in 1953, returning as a visiting professor to the California Institute of Technology (Caltech) where he had been an undergraduate. At Caltech Shockley met Arnold Orville Beckman (1900–2004) who held a Caltech Ph.D. and had founded a successful business, Beckman Instruments, on the basis of a pH meter he had invented in the 1930s. Beckman was specifically interested in the development of automation, rightly seeing the transistor as central to this process. Not surprisingly he was impressed with Shockley, whose reputation as a physicist and understanding of semiconductors were more or less unequalled, and he agreed to back a new business venture – Shockley Semiconductors. Although Beckman's preference was for the business to be located in southern California, in the vicinity of Caltech, Shockley insisted that it be established in northern California, in the area where he had grown up and where his mother was living in Palo Alto. The area was also adjacent to Stanford University where the provost, Frederick Terman (1900–82) was in the process of creating the world's first university-owned science park and was eager to collaborate with Shockley (Shurkin 2006). Between them Shockley and Terman were the founders of what became known as Silicon Valley.

Shockley Semiconductor laboratories opened in 1956 and Shockley recruited a team of young engineers and scientists, including Gordon Moore who held a Ph.D. in chemistry from Caltech, and Robert Noyce (1927–90), an engineering Ph.D. from MIT. Shockley was a genius but also something of a tyrant and soon Moore and Noyce, together with several other researchers, decided to break away. They made contact with the New York-based Fairchild Camera and Instrument Company which agreed to fund a new division devoted to semiconductor research. Fairchild Semiconductor opened in Mountain View, California in 1957 with Moore as the manager of engineering and Noyce as division manager. In 1963, Moore and Noyce were joined by Andrew Grove, an Hungarian refugee with a Ph.D. in chemical engineering from the University of California at Berkeley. Grove became the leading organizational force in Fairchild Semiconductor which, by 1967, had annual sales of US$130 million and a staff of 15,000. In 1965 Moore had stated in a journal article that the number of transistors on a microprocessor doubled every eighteen months to two years – a speculation which the Caltech professor and entrepreneur Carver Mead later dubbed 'Moore's Law'. Meanwhile, Shockley Semiconductors had failed; a victim of William Shockley's temperament, lack of business acumen and lack of a viable product. However, in spite of commercial success, Moore and Noyce became frustrated with being part of the Fairchild empire and in 1968 they broke away again, taking Andrew Grove with them, and founded Intel – derived from 'integrated' and 'electronics'. Intel's function was to investigate and then commercially develop the possibilities of semiconductor memory by continually cutting production costs while cramming ever more transistors on to a single chip.

In 1970 the company introduced its first successful product, the 1103 chip which contained 1K, or a thousand bytes, of dynamic random memory (DRAM). By 1971 the company's sales had reached almost US$9.5 million dollars and the company was able to raise almost US$7 million from an initial public offering of its shares. All of this, however, was merely a prelude to the technical breakthrough which would really propel the company's growth – the invention and introduction of the INTEL 4004 microprocessor (Gross 1996). The invention of the microprocessor generated the proliferation of personal computers as well as all manner of appliances ranging from digital watches to calculators. Intel grew to be an immense organization, continuing to spend massively on R&D and improved production techniques and facilities and becoming familiar to consumers through the presence of its logo and the tag line 'Intel Inside'.

BILL GATES AND MICROSOFT

By the 1990s the revolution in information technology was having a major impact in all the advanced economies. Mainframe computing had been joined by a huge growth market in personal computers (PCs) which first appeared in the early 1970s. In 1974 a small calculator company based in Albuquerque, New Mexico launched a machine called the Altair. The machine lacked a keyboard and a monitor and was mainly aimed at hobbyists. Bill Gates, then a 19-year-old Harvard undergraduate, was just such a hobbyist and he was quick to see the machine's potential. Together with his friend Paul Allen, Gates wrote a version of the existing BASIC programming language for the Altair and established the company which later became Microsoft (McCraw 2000). Gates and Allen relocated to Albuquerque and worked frantically on developing versions of FORTRAN and COBOL to run on the Altair. The technically limited Altair was, of course, doomed as larger companies began to invest in producing PCs. This made little difference to Gates who had already decided to concentrate on software development to the exclusion of everything else.

In 1977 the Tandy Corporation introduced the TRS-80 computer which, with a video display terminal and a keyboard, resembled later PCs. Microsoft licensed Tandy to install its version of the BASIC computer software in the TRS-80. This was a major deal for the company and Microsoft's sales in 1978 were approaching US$1.5 million. Gates relocated the company to his home town of Seattle – far away from the major centres of computer development in northern California and Massachusetts. However, Gates calculated that in the world of software development location was relatively unimportant – what counted were brains and computers. He established Microsoft in the Old National Bank building in Bellevue and drove forward – employing 130 staff and attempting to lead the infant company by example. When IBM finally decided to enter the PC market in the 1980s (Apple having already entered in 1977) they rapidly became the leading supplier but also made strategic mistakes which were later to cost them dearly. Eager to cash in on the PC boom as quickly as possible they decided to outsource both the microprocessor and the disk operating system. As a result Intel and Microsoft enjoyed a rich bonzanza and, because of IBM's prior move to an open architecture, the two companies were able to supply not just IBM but also to the makers of cheap IBM clones (McCraw 2000). As other companies entered the PC market they offered quicker delivery schedules and lower prices than the inflexible and highly bureaucratic IBM.

By the mid-1990s, when Microsoft launched the Windows 95 operating system with an advertising campaign only equalled in recent times by Apple, Bill Gates' personal fortune was calculated to be in the region of US$18 billion. Meanwhile IBM, having survived a decade of decline, had transformed itself into a services business with an emphasis on consulting and e-commerce. The initial encounters between IBM and Microsoft contained an element of culture shock when IBM's blue-suited, middle-aged executives were confronted with the young Microsoft dudes led by the youthful Bill Gates (Gross 1996). Gates' image and that of other emerging IT stars shattered the mould of what corporate man and woman's image should be. Further, the headquarters of the new generation of tech companies broke away from the accepted tower-block style and adopted a campus based approach which appeared to celebrate individuality rather than conformity.

CONCLUDING REMARKS

McDonald's, Intel and Microsoft have progressed from obscure start-ups to become huge global corporations. Their success, albeit contested, must be rooted in providing something that their customers want and, given that they are not monopoly providers, at a price their customers are willing to pay. However, they did not achieve all of this by staying quiet about themselves. McDonald's have always been enthusiastic advertisers and have worked hard to establish an instantly recognizable brand – from the golden arches to Ronald McDonald and the 'Happy

Meal'. According to Schlosser and Wilson, the McDonald's Corporation spends more on advertising than any other brand of food (Schlosser and Wilson 2006). In the early days of the company Ray Kroc was particularly inspired by Walt Disney (1901–66) with whom he had served in the US army during the First World War. Kroc particularly admired Disney's vision which combined technical progress in film animation with a magical, fantasy world aimed at children. Kroc similarly aimed his products at children, devising restaurant layouts with play areas calculated to be attractive to kids and therefore families. In all of this he was assisted by changing demographics. In the ten-year period following the Second World War the USA experienced a baby boom and the number of children increased by 50 per cent. This coincided with an economic boom, with jobs plentiful and spare money to spend. Cheap eating out became part of a family's way of life – instead of mom cooking Sunday lunch at home with all the fuss and bother involved, why not drive to McDonald's, have some fun food and get a free toy? In the fullness of time more moms became working mothers with less time (or inclination?) to cook kid's meals and so the era of the take-away developed. Seemingly, whichever way things went – whether it was eat-in, take-away or drive thru' – McDonald's was there to ride the wave. Ray Kroc knew about brand loyalty and the fact that loyalty formed in childhood can last a lifetime – the phrase 'give me the child until he is seven and I will give you the man' sums things up. We will examine aspects of marketing and advertising further in later chapters but, for the moment, it is sufficient to say that the McDonald's approach has succeeded in inserting the brand – the red and gold packaging, the product range, the sheer comforting familiarity – deep into modern society and, by the way, am I the only academic who really enjoys a Quarter Pounder meal with a Hot Apple Pie?

Meanwhile, Intel were late arrivals on the brand building and advertising scene, seeing themselves as somewhat above such things. It was only in the late 1980s that the company began to give serious thought to marketing its products beyond the tech-savvy world of design engineers working for personal computer companies, to the general public. This shift in strategy, stemming from the ideas of a single individual, provides a perfect example of innovation within a large, established corporation. Originally an engineer, Dennis Carter joined Intel in 1981 from the Harvard Business School where he had completed an MBA. Working as technical assistant to the company's chief executive officer Andrew Grove, Carter became aware that Intel faced major challenges in the rapidly maturing microprocessor market (Harnish 2012). The impact of 'Moore's Law' was to ensure constant technological obsolescence. For example Intel's 16-bit microprocessor (the 286) was overtaken in three years by the company's 32-bit microprocessor (the 386). However, demand for the new product was sticky, and the 386 was slow to be adopted even as a superior product – the 486 – was becoming feasible. Basically, technological improvement was running ahead of end-user awareness. Dennis Carter's proposed remedy for this was to develop and launch a marketing campaign aimed at raising awareness of Intel's microprocessors among the general public. In

1988 he persuaded Andrew Grove to fund a pilot campaign aimed at establishing the Intel brand in the minds of consumers – essentially giving the company's product an identity. The success of this initial campaign, which among other things succeeded in eclipsing the 286 and establishing the 386, led to a major marketing campaign calculated to coincide with the launch of the 486. Interestingly, a 1991 court ruling determined that Intel were unable to trademark mere numbers like 286, 386 or 486. This led the company to search for an umbrella brand and prompted Carter to approach the PC manufacturers with the offer of a co-operative marketing scheme, whereby Intel would pay 3 per cent of its revenues from microprocessor sales into a fund that would pay half of a PC manufacturer's advertising costs – as long as the ad concerned promoted Intel-based PCs carrying Intel's logo. Dell signed on immediately for the scheme and, by the end of 1991, 300 manufacturers had agreed to join – perhaps most famously IBM. IBM's advertisement in *The Wall Street Journal* was the first to give equal billing to the 'Intel Inside' tag line and logo. By the end of 1992, Intel's worldwide sales had increased by 63 per cent and by 1995 brand awareness of Intel among European consumers had risen from 24 to 94 per cent. Meanwhile, a three-second jingle had been added and the Intel Inside slogan supplemented by the Pentium brand. By the time Carter retired in 2000 Intel had become one of the most valuable brands in the world.

Finally, a few remarks about the role of *venture capital* – self-evidently capital is an essential ingredient in entrepreneurial activity and the development of innovation. An obvious source of capital is the entrepreneur's personal assets. In the case of Intel, for example, Moore and Noyce made their original contact with Fairchild Camera and Instrument Company through a San Francisco-based investment banker called Arthur Rock. Fairchild Semiconductor was funded using Rockefeller money invested by what would later become the venture capital vehicle, Venrock Associates. When Moore and Noyce wanted to break with Fairchild they consulted Rock again and he succeeded in raising US$2.5 million to invest alongside their own contribution of US$500,000. In the case of Thomas Edison, the commercialization of his inventions was funded by specific arrangements with bankers such as J.P. Morgan (1837–1913). This is so-called 'angel' or seed funding of the kind which might support the staging of a Broadway show, a movie, or the initial sponsoring of a rock band. The investor is looking for returns on the sum invested but the relationship is personal rather than corporate. Venture capital, however, is a subset of private equity and seeks to match institutional investors with potentially highly profitable ventures. The so-called founder of the venture capital model was George Doriot (1899–1987), a former Dean of the Harvard Business School and founder of INSEAD (European Institute of Business Administration) at Fontainbleu near Paris. Doriot was instrumental in establishing the American Research and Development Corporation (ARDC) in 1946. In 1957 ARDC invested US$70,000 of its client's money in a company called Digital Equipment Corporation which was valued at US$355 million following the company's initial public offering (IPO) in 1968 – an annualized rate of return of 101 per cent. Venture capital was prominent in the funding of many of the enterprises in Silicon Valley and remains closely associated with the so-called high-tech industries – industries which often carry a very substantial level of risk (Ante 2008). Henry Kressel, who successfully made the transition from the technical world at RCA Labs to the world of venture capital has noted that, like Intel, many of today's giant companies including Oracle, Cisco Systems, Apple, eBay and Google were all initially venture-backed start-ups. However, he also cautions that although some ventures succeeded, many others failed. Funding of early-stage technology ventures is therefore very risky and for this reason venture capital investors demand high rates of return on the funds committed (Kressel 2010).

CHAPTER 5

INDUSTRIAL DESIGN AND INNOVATION

INTRODUCTION

This chapter examines the developing role of the industrial designer and explores the links between design and product innovation. Modern industrial design has its roots in the nineteenth century and specifically the rationalization and standardization of products which came initially with the American system of manufacturing and, later, mass production. The catalyst for the widespread awareness of developments in the USA was the Great Exhibition of 1851 held in London's Hyde Park. By the middle of the nineteenth century the USA had begun the process of large-scale production of standardized products, using interchangeable parts and deploying power machine-tools set in sequence, which would ultimately evolve into mass production (Heskett 1980). For example, in 1855, Samuel Colt (1814–62) opened the Colt Armory in Hertford, Connecticut for the bulk production of revolvers, employing the engineer Elisha K. Root (1808–65) to tool-up the machine shops. Root installed a range of drop hammers, boring and milling machines and also devised the necessary jigs and gauges to achieve greater accuracy. An indication of the growing reputation of the US machine tool industry can be seen in the decision by the British government in 1856 to install American equipment driven by steam engines in the new machine shop at the Royal Small Arms Factory in Enfield north of London (Habakkuk 1962). Greater accuracy in machining substantially reduced the amount of hand-finishing of components required to enable final assembly of the product to occur. Levels of accuracy in engineering were further improved by Joseph R. Brown's (1810–76) invention of the Vernier caliper in 1851 and the introduction in 1868 of the pocket micrometer. These instruments, capable of measuring one-thousandth of an inch and available at a price affordable by the average machinist, greatly enhanced the emergence of precision engineering. By the time Henry Ford was putting the Model T Ford into mass production, the quality of components had become sufficiently reliable to feed the moving assembly track in the quantities essential to produce a volume car, which leads us to a consideration of the nature of design itself.

In his book *Objects of Desire: Design and Society Since 1750*, Adrian Forty (2000) provides two common uses of the word design. The first merely refers to the way things look, whether an object is aesthetically pleasing, and involves judgements of taste. The second, which Forty considers more exact, is concerned with the development of instructions for the production of manufactured goods. According to Forty the two meanings are in fact inseparable – the way things look being intimately connected to the conditions of their making. In the case of Henry Ford he was interested in design in the second sense offered by Forty but, certainly in the early days, saw issues of image as secondary (Brinkley 2004). The technical aspects were paramount and utility always prevailed over styling. Nevertheless, even the rather functional-looking Model T played its part in defining what a car should look like. This leads on to a further point made by Forty, namely the ways in which new innovations are made acceptable/desirable to potential consumers through the mediation of design. Increasingly, the technological realities of the product are disguised by external styling. Forty illustrates his point using the example of the design of early radio cabinets which developed during the 1920s and 30s from crude assemblies of resistors, wires and valves via imitations of antique furniture to modern forms (initially of plywood but later moulded from Bakelite) suggesting a product belonging to the future.

THE INFLUENCE OF MORRIS, BEHRENS AND LOEWY

William Morris (1834–96) was a visionary socialist, poet and writer as well as an accomplished craftsman and designer. His influence on modern design was enormous which is somewhat ironic – considering that he despised the modern world and detested what he saw as the cheap, degraded products produced by modern industry. Morris was opposed to the extreme division of labour which characterized modern production methods. Instead, he took inspiration from the Middle Ages, a period he imagined as being characterized by craftsmanship, respect for materials and social harmony. He was not alone in taking this view which was part of an identifiable phenomenon in British nineteenth-century thought, ranging from the architectural designs of Augustus Welby Pugin (1812–52), the aesthetic criticism of John Ruskin (1819–1900), the later novels of Charles Dickens (1812–70) and the rural ramblings of Richard Jefferies (1848–87). Martin Weiner has convincingly associated this phenomenon with what he terms the decline of the industrial spirit – a turning away from the realities of modern industrial society in search of some illusory rural idyll (Weiner 2004). Pugin and Ruskin inspired what became known as the Arts and Crafts Movement – a group, including Morris, which embraced traditional craft methods, vernacular styles of architecture and simplicity rather than sophistication in design. In the case of William Morris the eventual conduit for making and marketing his products was the firm of Morris & Co. founded in 1875 and offering a range of furniture and textiles, including many designs which remain available today. Interestingly, given Morris's political views, his commitment to craft-based, labour-intensive production methods meant that the bulk of the population were unable to afford his products (Heskett 1980).

The Arts and Crafts Movement tended to celebrate the simple life and the idea of community on a quasi-religious basis, electing to retreat into the countryside and work on a basis inspired by the medieval Guilds – an example being the Guild of Handicraft established at Chipping Camden in Gloucestershire in 1902 by Charles Robert Ashbee (1863–1942). Ashbee's influence was significant in diffusing the ideas of the Arts and Crafts Movement to Europe and the USA and specifically in the establishment in 1907 of the Deutscher Werkbund (DWB). In contrast to Morris and other earlier members of the Arts and Crafts Movement, Ashbee was willing to contemplate compromising some of the movement's principles by accepting the necessity of standardization in order to facilitate mass production (Woodham 2004). Also, whereas Morris and his associates tended to be insular in their outlook, Ashbee took a broader view, exhibiting his work in Austria, France and Germany and travelling to the United States where he met the influential architect and designer Frank Lloyd Wright (1867–1959).

The DWB, although partly inspired by the ideas of the Arts and Crafts Movement, had a different agenda. Rather than turning away from the mainstream of economic and industrial development, it sought instead to improve the standards of contemporary German industrial design by bringing together designers, manufacturers and writers in a single, collaborative organization – basically an attempt to combine efficiency in production with the creation of aesthetically pleasing artefacts. Perhaps the most influential member of the DWB was Peter Behrens (1868–1940). Behrens worked as the artistic director of the leading German electrical firm AEG from 1907 to 1914, designing everything from arc lamps, fans, kettles and clocks to publicity material, the company logo, the company typeface and industrial buildings such as the famous Turbine Hall in Berlin. Behrens designs were innovatory, in keeping with an innovative and essentially modern industry, and also provided AEG with perhaps the first example of a corporate identity (Woodham 2004). Behrens' work for AEG in many ways anticipated the approach taken by American designers in the 1920s and 30s of which Raymond Loewy (1893–1986) was the most prominent, even appearing on the cover of *Time* magazine in 1949 surrounded by many of the products he had designed.

Loewy was unashamedly commercial describing good design as an upward sales curve and producing numerous company logos (e.g. for Shell). He was strongly influenced by the fashion for streamlining during the 1930s, successfully exploiting the link in the minds of consumers between speed and modernity – even in static products such as refrigerators and pencil sharpeners (Tretriack 1999). The success and publicity gained by Loewy and other designers of his era prompted manufacturers to give greater recognition to the competitive advantage offered by good design (Meikle 2005). Further, a greater understanding of ergonomics, also known as human engineering or human factors in the USA, began to influence design standards. The scientific managers such as Taylor, Gantt and particularly the Gilbreths had been aware of issues relating, for example, to posture and fatigue as early as the 1900s (Sheldrake 2003). By the 1960s ergonomics had become increasingly scientific and been joined by anthropometrics, the systematic study and measurement of the human form, when designers were concerned with designs involving human movement. Most prominent in the field was Henry Dreyfuss (1904–72) who published his influential books *Designing for People* in 1955 and *Measure of Man: Human Factors in Design* in 1959. During the twentieth century the number of white-collar jobs grew steadily in the USA as organizations became increasingly complex. When the sociologist C. Wright Mills (1916–62) published his book *White Collar* in 1951 he commented on the transformation of office work as new technologies eroded former patterns of employment, creating new jobs but rendering others redundant. By the end of the twentieth century, evolutionary change became revolutionary change with the introduction of computers. This leads us to a consideration of the role of design at a company which was for many years central to many of these changes – International Business Machines (IBM).

ELIOT NOYES AND DESIGN AT IBM

The first modern electronic computer used in business was the Univac, introduced by Remington Rand in 1950. However, it was IBM which rapidly emerged as the dominant force in the industry and it was under IBM's leadership that information technology became the most dynamic industry of the late twentieth century (McCraw 2000). Good industrial design, together with a strong corporate identity, was central to IBM's success. The leading figure in this development was the Harvard trained architect Eliot Noyes (1910–72). Noyes was strongly influenced by leading European modernist architects such as Le Corbusier (1887–1965) and Walter Gropius (1883–1969), both of whom had worked for Peter Behrens in their early careers (Woodham 2004). Noyes spent a year with Gropius before spending several years as director of industrial design at the Museum of Modern Art (MOMA) in New York and later working for the leading designer Norman Bel Geddes (1893–1958). He developed a close working relationship with the head of IBM, Tom J. Watson Junior (1914–93) and when Watson became convinced that IBM required a corporate design programme, he turned to Noyes to lead it. Noyes became the company's Consultant Director of Design with very wide powers and set out to provide IBM with an elite image in keeping with its corporate culture, producing a set of IBM standards and establishing design specifications for every product (Heskett 1980).

At IBM design became closely linked with technical innovation. In 1961 the company introduced the 'Selectric' typewriter which typed by the novel means of a so-called golf-ball – a moulded sphere carrying the various characters which moved across the paper, shifting and turning to present the appropriate aspect. Noyce designed a similarly radical casing which emphasized the elimination of the type-bars and movable carriage. As IBM developed its computer systems during the 1960s, Noyes produced designs to match. In 1964, after conducting what McCraw (2000) has described as the most expensive privately financed R&D effort in American business history, the company introduced the 'System 360' computer system. Small, compact and very versatile it consisted of a series of rectangular, modular units. Some of the working parts were left visible and the console was designed in co-operation with the company's ergonomics section to be comprehensible and easily operated. The System 360 was a huge success and rendered IBM so strong that the company emerged as the natural market leader in information technology for the next 30 years. Appropriately, Noyes' designs made use of the new plastics, contributing to the system's image of efficiency and cool modernity and also enhancing IBM's reputation for high-quality design as well as technical sophistication.

DIETER RAMS AT BRAUN AND WHAT CONSTITUTES GOOD INDUSTRIAL DESIGN

The German electrical consumer goods company Braun has become famous for the quality of its designs and has used them to establish a strong corporate identity. The firm was established in 1921 by Max Braun (1890–1951) in Frankfurt, initially to manufacture radio components but subsequently producing entire radios including the first battery-powered portable radio in 1935. When Max Braun died his sons Artur and Erwin Braun took over the firm, extended the product range to include shavers and, in 1957, the famous Multimix food processor. The brothers were interested in modern design and they commissioned the industrial designer Wilhelm Wagenfeld (1900–90) to provide fresh designs for the Braun range of radios and record players. Wagenfeld had been associated with the famous Bauhaus which, inspired by the Arts and Crafts Movement and led by Walter Gropius, had pioneered modern design during the 1920s and early 30s before being closed by the Nazis in 1933. They also appointed Dr Fritz Eichler, a lecturer at the Ulm School of Design, as head of design and he began to develop the cool, minimalist approach which was to become the hallmark of Braun products (Heskett 1980).

In 1955 Eichler was joined by Dieter Rams and the characteristic 'Braun look' began to emerge, initially embodied in the 'Kitchen Machine KM 321' of 1957 and subsequently applied to a wide range of domestic products including radios, record players, clocks and electric razors. Many of Dieter Rams' designs became collectibles – MOMA for example began to collect Braun products as early as 1958. Rams reputation extended far beyond the narrow confines of often anonymous corporate design and his influence has been substantial – not least Sir Jony Ive, the lead designer for Apple, has acknowledged Rams' influence. During the 1970s Rams became increasingly concerned about the state of the world and the impact of the 'throw away' mentality. His response was to develop what he described as his *ten principles of good design* as follows: good design is innovative, good design makes a product useful, good design is aesthetic, good design makes a product understandable, good design is unobtrusive, good design is honest, good design is long-lasting, good design is thorough down to the last detail, good design is environmentally friendly, good design is as little design as possible. Note the contrast with what Rams saw as the excess displayed by some American designers and also the influence of the minimalist notion of less is more. Rams might have added that good design should always be executed with manufacturability and unit cost in mind. Braun's total commitment to quality design has not, however, always led to commercial success. In 1967 the company was taken over by Gillette whose influence resulted in a rather more colourful, less restrained approach to design (Woodham 2004). Braun is currently owned by Procter & Gamble.

FRANK PICK AND LONDON'S TRANSPORT NETWORK

Frank Pick (1878–1941) presided over the development of London's underground and bus network during the first half of the twentieth century. Pick studied law at London University, qualiying as a solicitor in 1902, before joining the North Eastern Railway company as a management traineee. He became a personal assistant to the company's general manager Sir George Gibb (1850–1925) and, when Gibb was appointed chairman of the Underground Group in 1906, Pick went with him. In 1908 Pick, having complained to Gibb about the Group's poor marketing strategy, was appointed company publicity officer. Pick decided to adopt the successful approach taken by the North Eastern and other railway companys to advertise their services using posters. In spite of having no previous experience in marketing, Pick began to commission posters exhorting passengers to use the Underground, not just as the means to get to work, but for all kinds of leisure activities – trips to London's suburban fringe, sporting events, parks, museums, theatres and so on – all calculated to increase passenger numbers. Pick reserved specific advantageous spaces at Underground stations for his posters and began the process of standardizing the format.

In 1916 he commissioned the calligrapher Edward Johnston (1872–1944) to design a specific typeface for the Undeground which resulted in the creation of what became known as 'Johnston Sans' – a variation of which, 'New Johnston', remains in use on the London Underground today. Pick also comissioned Johnston to redesign the company symbol, which resulted in the creation of the famous red roundel with the station name on a dark blue bar. Pick introduced the roundel and Johnston Sans lettering across the entire network, thereby giving London Undergound the basis for a recognizable corporate identity. In the early days Pick was content to have the posters designed by anonymous commercial artists. However, as his ideas developed he began to commission leading modern artists to produce posters and even to hold exhibitions of modern art and design at London Underground stations. Man Ray (1890–1977) and Graham Sutherland (1903–80) were among the prominent artists commissioned by Pick but perhaps the most outstanding works were those created by Edward Mc Knight Kauffer (1899–1954), whose output for London Underground during the 1930s included posters in styles ranging from Cubism to Surrealism (Ovenden 2013).

In 1933 London's five privately owned underground railway companies, together with the municipally and privately owned tramway and bus companies, were taken into public ownership and merged to form the London Passenger Transport Board. Frank Pick was appointed managing director and tasked with creating a modern, inegrated transport network for London to operate alongside the rail network. This massive task gave Pick the opportunity to apply his ideas on design to the building of a corporate headquarters at 55 Broadway above St James's Park Underground Station and, as the underground system expanded into the suburbs, the construction of numerous stations – many of which are now listed buildings. Pick's interests encompassed every aspect of design from humble bus stops to station interiors and the fabric designs for the seats of buses and underground trains. Among the practical problems Pick encountered was that of enabling passengers to successfully navigate the network. To assist in this task he used a schematic developed originally in 1931 by Harry Beck (1902–74), a London Underground draughtsman. Appropriately, Beck based his map on an electric circuit, representing each specific line in a different colour and marking interchange stations as diamonds. The central, more congested, area was enlarged for legibility and the route of each line simplified by using only verticals, horizontals and diagonals. Beck's map was introduced in 1933 and remains the basis of today's map – not just in London but also in many other cities including Sydney and New York. When the Design and Industries Association (DIA) was founded in 1915, Frank Pick became an inaugural member. Inspired by the Deutscher Werkbund, the DIA sought to improve the quality of design in British industry by promoting better understanding between manufacturers, designers and retailers. In 1934 Pick was appointed chairman of the Council for Art and Industry established by the British government to improve public understanding of the social cultural and aesthetic benefits of good design (Woodham 2004).

THE RISE OF RETRO

L.P. Hartley's (1895–1972) 1953 novel *The Go Between* begins with the famous line 'The past is a foreign country; they do things differently there.' In his book *Retromania*, Simon Reynolds (2011) paraphrases this statement to offer an explanation for the rise of retro, claiming that economic shifts, technological innovations and sociological changes generate increasing nostalgia for the past, however recent, until it is the present rather than the past which becomes a foreign country. Retro differs from heritage in that it relates to the recent past and often to ephemera in music, fashion and style generally – tending to value mass culture rather than folk or high culture. Although the word retro is of recent origin, having first been used in France in the early 1970s as an abbreviation of retrograde, the phenomenon itself is far older (Woodham 2004). For example, the English author and poet Thomas Hardy (1840–1928) spoke of reliving an earlier, possibly erotic experience, in his 1913 poem *At Castle Boterel*. In Hardy's case an event of over 40 years earlier was strongly evoked by revisiting a specific place and landscape. In the case of the French writer Marcel Proust (1871–1922) it is the taste of a small cake (a madeleine) soaked in lime blossom tea (tisane) that unlocks the involuntary memories of his narrator's previous life in his immense novel *In Search of Lost Time*. However, Hardy and Proust are perhaps a little too highbrow for modern retro. Closer to the current meaning is a pop song called *Our Favourite Melodies* first recorded in the USA by Gary Criss and covered by the English pop star Craig Douglas in 1962. In the song the singer addresses his former girlfriend, complaining that wherever he goes he hears the songs that were especially theirs, including Ray Charles's (1930–2004) *Hit the Road Jack*; James Darren's *Goodbye Cruel World* and Bobby Vee's *Run to Him* – all big hits in 1961 – causing him to relive happier times. A similar approach is taken in The Carpenters' 1973 hit *Yesterday Once More* which deals directly with the power of music to transport us back to a specific time and experience – a song rendered all the more poignant given Karen Carpenter's (1950–83) death at a tragically young age as a result of anorexia nervosa.

According to Reynolds, retro is always concerned with things which existed in living memory rather than some far-off period such as the Middle Ages which, as we have seen, inspired William Morris. Also, retro involves what Reynolds describes as exact recall, sustained by the ready availability of archived material, whether it is documents, photographs, music, recordings or fashion. In this context the Internet has been particularly influential, facilitating access to vast stores of material otherwise difficult or impossible to obtain. In the case of pop music, for example, it now has its own history dating back over 60 years with most of the material available online. Further, retro is sourced from charity shops and flea markets rather than auction houses and antique dealers (Reynolds 2011). Having said all of this, retro now has an influence which transcends the subcultures of the street. As the boundaries between high culture and popular culture have become blurred so the worlds of the museum, the vintage clothes shop, the tribute band and, so called, *retro-futurism* have converged. Pastiche products are now created which deploy state-of-the-art technology while being strongly evocative of the past – see for example the work of J. Mays, currently the chief creative officer for the Ford Motor Company and designer of the Volkswagen New Beetle (Meikle 2005).

HARLEY-DAVIDSON MOTORCYCLES

Harley-Davidson is arguably the world's best-known motorcycle brand – certainly in the sphere of retro. The company was officially established in 1907 by William S. Harley (1880–1943) and Arthur Davidson (1881–1950) in Milwaukee, Wisconsin and, by 1910, was producing 3,000 bikes each year. During the First World War the company was contracted to supply 20,000 bikes for the US Army and, by the 1920s, had become the USA's largest motorcycle manufacturer with a nationwide network of dealerships and a growing export market (Dunne 2012). Famous for their size (all modern Harleys are over 750cc), design and distinctive exhaust note, Harley-Davidson bikes have featured in numerous films, particularly road movies, and the strength of the brand is evident in associated clothing and fragrances. In spite of all this, however, the company almost melted down during the 1970s as a combination of recession, Japanese competition and poor quality took it to the brink of bankruptcy.

In 1981 a group of executives, including design chief William G. Davidson, the grandson of the joint-founder, gained control of the company from American Metal Foundries, its corporate parent. Crucially they persuaded the US government to impose temporary protectionist tariffs on Japanese imported motorcycles. This gave the company a breathing space while they totally transformed their methods of production, adopting the just-in-time and total quality management techniques (see Chapter 6) deployed by their Japanese competitors and removing the traditional distinctions between blue- and white-collar workers. The new owners also decided to robustly promote Harley-Davidson as an American product which was proud of its heritage. As quality and reliability improved Harley succeeded in moving their products up market, producing fewer bikes but selling them at a higher price. In 1984 the company introduced the Evolution big twin engine, which powered a new generation of bikes including the Super Glide touring machine. By the late 1980s a high-end motorcycle had become a status symbol for the yuppie generation – a generation which had been captivated by Robert M. Persig's 1974 novel *Zen and the Art of Motorcycle Maintenance: An Inquiry into Values*. Whereas the company had traditionally drawn its customers from the young working class, by 1990 the typical buyer was a 35-year-old male in a professional occupation earning far above the national average and in the majority of cases holding a degree (Gross 1996). As its financial position improved, the company released a range of models which celebrated Harley's heritage, blending retro styling developed by 'Willie G' Davidson with leading-edge technology such as the liquid-cooled, fuel-injected, V-twin Revolution engine introduced in 2001. A combination of sound management and technical innovation has so far preserved an iconic brand.

CONCLUDING REMARKS

Industrial design currently occupies a central role in the process of producing and marketing successful products. In mature markets it is essential that manufacturability is matched by aesthetic desirability. Consumers have become ever-more aware of design, markets have become ever-more segmented, and novel materials and innovative production methods have stimulated the phenomenon known as mass customization. In his book *Future Perfect* the business thinker Stan Davis (1987) reflected on the ways in which new technologies would enable companies to better serve consumers. Briefly, he argued that as markets mature consumer expecations evolve and the one size fits all approach becomes less viable. New technologies provide the possibility of a far more nuanced approach to production, creating business opportunities which derive from expanding consumer choice (however trivial some of this choice might seem). The underpinnings of mass production had been standardized products produced in vast quantities and aimed at homogeneous markets – Ford's Model T was the paradigm example. By the final years of the twentieth century, however, this approach had been undermined by numerous technical and demographic changes and also the Japanese system of lean production which was specifically created to compete with American mass producers – achieving lower costs at smaller volumes with higher quality (Pine 1993). Mass customization is a synthesis of two competing systems of management – the mass production of individually customized goods and services. In mass production low costs are achieved through economies of scale – reducing unit costs of a single product through greater output and faster throughput. In mass customization low costs are achieved through economies of scope – using a single process to produce a variety of products more cheaply and more quickly. Successful mass customization companies achieve both, gaining economies of scale on standard components and then combining them in numerous ways to provide economies of scope. In this situtation design becomes ever-more significant in terms both of manfacturability and in better meeting consumers' preferences. Meanwhile, the rise of retro has stimulated an interest in products which, while not being absolute replicas of former designs, allude to them in a recognizable and meaningful way.

CHAPTER 6

IMPACTS OF TECHNOLOGICAL CHANGE

INTRODUCTION

This chapter reflects on some of the impacts of technological change through an examination of developments in aspects of manufacturing, automotive engineering, containerization and computer applications, including the Internet. Computers have become ever-more significant, increasingly intruding their way into how we do things and even how we think. For example, the development of word processing ended the era of the typewriter, closed the typing pools and made the job of the typist redundant. In newspaper printing a total revolution in production techniques occurred in the 1980s as a result of which new computer-based technology rendered many traditional skills redundant. In the sphere of recorded sound the 78 rpm vinyl disc was replaced by the 45 rpm during the 1960s, while vinyl discs were increasingly replaced by cassette tapes during the 1970s and these were replaced in turn by compact discs during the 1980s. For a period in the 1980s the Sony Walkman portable cassette player was the must-have item (Morita 1986), radically changing consumers' expectations about the delivery of music and ultimately selling 150 million units. Although vinyl discs, cassette tapes and compact discs continue to be available, the whole market was revolutionized in the 2000s with the introduction of computer-based digital downloads. In the world of manufacturing a series of piecemeal changes occurred leading to the emergence of what is currently described as advanced manufacturing.

Beginning in the 1960s, the elements of advanced manufacturing were introduced gradually as materials and technology evolved – an example being the industrial robot. In the 1950s George Devol (1912–2011) designed a mechanical arm which could be programmed to repeat precise tasks such as grasping and lifting. He applied for a patent in 1954, naming the concept Universal Automation – later shortened to Unimation. Joseph Engelberger licensed Devol's patent, formed a company Unimation Inc., and produced the Unimate, an early and highly successful attempt to replace factory workers with robotic machinery. In 1961 General Motors put the first Unimate arm on an assembly line at their plant in Ewing, New Jersey, the device being used to lift and stack hot die-cast components direct from their moulds. Both Chrysler and Ford followed suit and soon Unimates designed for many potentially hazardous production jobs were being manufactured. Japanese automotive manufacturers were early adopters of the Unimation vision of mobile, remotely controlled robots and rapidly overtook the USA in terms of numbers of robots deployed.

TECHNOLOGICAL INNOVATION AND THE QUALITY MOVEMENT IN JAPAN

The Japanese enthusiasm for robotics was linked to the the emerging quality control movement in that country's industrial sector, particularly automitive engineering. As it happens, the quality control movement in Japan was largely inspired by American thinking, particularly the ideas of W. Edwards Deming (1900–93) and Joseph M. Juran (1904–2008). In the years following the Second World War Japanese products had a reputation for indifferent workmanship and poor quality. An effort to improve the situation was spearheaded by American production engineers operating under the auspices of the Civilian Communications Section (CCS) of the Occupation Administration. Two of these engineers were Charles Protzman and Homer Sarasohn (1916–2001) who wrote a complete course on industrial management in which they emphasized the significance of quality control, quoting the industrialist and steel maganate Andrew Carnegie's (1835–1919) famous dictum – 'there lies still at the root of great business success, the very much more important factor of quality'.

During the final months of 1949 and the early months of 1950 Protzman and Sarasohn conducted two eight-week seminars in Tokyo and Osaka to which only top company managers were invited. Attendance at these seminars was compulsory for the participants and even when the Americans ceased running them the Japanese took them over and continued for the next 25 years. It was the participants in the CCS seminars who were active in inviting W. Edwards Deming to Japan in 1950, thereby initiating the Quality Control Circle movement. Interestingly, Protzman had been an engineer at the Hawthorne Works of the Bell System's Western Electric Company where the famous Hawthorne experiments in industrial relations were conducted between 1924 and 1933. These experiments, an attempt to extend the research ethos of Bell Labs to the world of work, were conducted at a time when major employers were becoming sceptical regarding the benefits of scientific management and beginning to temper their approach with insights from the academic world (Gillespie 1993). Beginning as a set of illumination studies examining the impact of various levels of lighting intensity on worker productivity, the Hawthorne experiments expanded to embrace a series of studies in worker–management relationships which eventually formed the basis of a new school of management thinking – namely, the human relations school. The most influential interpretation of the experiments was that put forward (one might almost say imposed) by Elton Mayo (1880–1949) and his followers and emphasized the benefits of a combination of benign supervision and the active nurturing of informal work groups (Sheldrake 2003). Protzman, however, was deeply sceptical of this interpretation, viewing the human relations approach as a prime example of what not to do. Instead, in the courses he prepared for his Japanese students, he placed the emphasis on old-style leadership on the model advocated by Henri Fayol. He also stressed the value of discipline, teamwork and cooperation between managers and workers, a message which struck a positive chord with his Japanese students who valued the national tradition of courage, selfsacrifice and benevolence.

Like Protzman (and indeed Joseph Juran) W. Edwards Deming was had once been employed at the Hawthorne Works of Western Electric. Deming trained as an electrical engineer at the University of Wyoming and gained a Ph.D. in mathematical physics from Yale. During his period at the Hawthorne Works he became familiar with the ideas of Walter A. Shewhart (1891–1967), the founder of statistical quality control, and in the early 1940s had established courses to teach Shewhart's methods to industrialists and engineers. It was these methods which he began teaching when he was invited to Japan in 1950. Deming gave hundreds of lectures to Japanese managers emphasizing the vital importance of statistical quality control. As a result 'Deming's Wheel', a diagram rationalizing the business of production management, was widely adopted. The interest Deming stimulated in the subject of quality control prompted the Japanese to invite Joseph Juran to Tokyo in 1953. Juran had worked with Shewhart and in 1951 had produced *The Quality Control Handbook*, the first such manual of its kind. Together Deming and Juran inspired the Japanese quality control movement which was eventually to be emulated throughout the industrialized world.

Toyota was one of the companies which adopted Deming's ideas. In 1961 the company was forced to withdraw from the US market as a result of low sales and poor-quality vehicles. However, by the time they returned in 1965, with the successful Corona sedan, the company had completely changed its approach to design and manufacturing on the pattern advocated by Deming (Harnish 2012). Total quality management formed a central factor in what became known as the lean production system in automotive engineering. In their best-selling book *The Machine that Changed the World*, Womack, Jones and Roos (1990) identified the principles of lean production. Based on the research findings of a worldwide survey of the automobile industry, the book traced the development of automobile manufacture from the early mass production techniques pioneered by Henry Ford, via the hybrid production system devised by Volvo, to the latest methods of lean production deployed by the Toyota Corporation – a transition from just-in-case to just-in-time which offered the possibility of major productivity gains.

According to Womack, Roos and Jones lean production consists of the following elements: integrated single piece production flow, with small batches made just in time; an emphasis on defect prevention rather than rectification; production which is pulled by consumption, not pushed to suit machine loading; team-based work organization, with flexible multi-skilled operatives; active involvement in problem-solving activities by all personnel, eliminating waste, interruptions and variability; and close integration of the whole supply chain from raw materials to retailing and distribution – all backed up by close buyer–seller relationships based on trust and co-operation rather than competition. In spite of its inherent fragility, lean production was soon adopted by major Western manufacturers and, during the 1990s, revolutionized automobile production by radically changing expectations concerning levels of both quantity and quality. Not surprisingly, perhaps, the ghosts of the scientific managers can be glimpsed in many aspects of lean production. However, the determination to get things right and keep them right is emphasized to a far greater degree – as is the awareness of supplier/customer focus. Insights from lean production, particularly the elimination of waste (*muda*) and the search for continuous improvement (*kaizen*), soon began to influence management techniques in many other industries. Finally, automotive engineering shifted from being a labour-intensive industry characterized by poor labour relations and endemic worker unrest to become a highly sophisticated capital-intensive industry, increasingly dominated by artificial intelligence.

MALCOLM MCLEAN, KEITH TANTLINGER AND CONTAINERIZATION

In the years between the 1950s and the early 2000s, world trade grew on an average of 5–6 per cent per annum. However, between 1990 and 2000, the rate of growth accelerated to 8–9 per cent, and from 2000 to 2008 it doubled again to 17–18 per cent, with trade between Asia and the rest of the world representing the biggest change. Much of this increase was facilitated by a simple box or shipping container – Jacoby estimates (2009) that 108 million cargo containers move worldwide each year and 90 per cent of manufactured goods move by container. This huge industry which revolutionized freight handling, has very simple origins in the 1950s with a North Carolina trucking magnate called Malcolm McLean (1913–2001). McLean was looking for ways to make his trucking operation more competitive and hit upon the idea of transporting his vehicles by ships along the Atlantic coast from North Carolina to New York. His original plan was to build roll-on-roll-off vessels which would substantially reduce transport costs. However, the cargo space lost by having to accommodate the actual vehicles, known as broken stowage, reduced potential cost-effectiveness to the extent that McLean modified his plan and decided to find a means of loading only the containers or boxes on to ships. In order to devise the technical means to make his idea a reality, McLean turned to Keith Tantlinger (1919–2011), the vice-president of engineering at a company called Brown Industries, which supplied McLean with truck trailers. Of course, there was nothing particularly novel in the idea of transporting cargos in metal boxes which had been used in shipping and on the railways since the nineteenth century to transport heavy loads such as coal. Where McLean's plan differed was in using larger, standardized containers and Tantlinger's major contribution was in designing the necessary twist-locks, corner posts, cell guides and spreader bars required to lift and lower the containers on and off ships and stack them safely. The twist-locks allowed containers to be secured with sufficient stability to enable the building of tall stacks, with obvious gains in freight capacity (Levinson 2006). At the time McLean and Tantlinger were developing their ideas, US regulations prevented trucking companies from owning shipping lines. In 1955 McLean therefore decided to sell his trucking operation, purchase a shipping company and obtain two former Second World War tankers which he had converted to carry containers on and below deck. He also began the process of designing a trailer chassis to carry removable containers. In October 1957 McLean's first container ship, *Gateway City*, was prepared to take the first shipload of cargo from Newark, New Jersey to Miami, Florida. Although containers had previously been transported by sea, there was no reliable knowledge of how a container vessel would perform at sea when fully loaded, the main fear being that the stacks might shift and create dangerous instability. In the event, when *Gateway City* docked in Miami after her four-day voyage, the container stacks had shifted by little more than one-quarter of an inch.

The use of containers massively reduced the time required to load and unload ships and, as an additional benefit, more or less prevented the leakage of cargo through pilfering. From relatively small beginnings containerization revolutionized the dock industry. In the Port of London for example, the upriver docks which had once employed thousands of men were closed, and cargo handling moved many miles down-river to capital-intensive container terminals. A similar pattern occurred globally as waterfronts were abandoned and new patterns of transportation established. By the 2000s huge container ports had been constructed, becoming the hubs of extensive transportation networks. Container vessels had become huge behemoths, up to 1,100 feet long and 140 feet across, capable of carrying 3,000 40-foot containers filled with 100,000 tons of goods and requiring a crew of as few as 20 (Levinson 2006). Whereas loading ships was once a slow, laborious and often dangerous process, a skilled crane driver can now lift a 20-ton container from a chassis and load it on to a container ship in a matter of seconds. A similar situation applies to unloading and the amount of time ships spend in port has been reduced to hours rather than days or even weeks. Containerization achieved for shipping what Henry Ford's moving assembly track did for automobile production, providing the basis for a standardized, integrated flow process. By reducing freight costs to a tiny fraction of production costs, containerization had a major impact on world trade, making goods produced in the low cost economies of Asia relatively cheap – benefiting Western consumers at the expense of Western manufacturers.

STEVE JOBS AND APPLE INC.

Reduced transport costs and the opening up of cheap labour markets in the Far East prompted the development of a new business model based on globalization. Whereas Henry Ford favoured the idea of concentrating as much as possible of the production process on a single site, this approach became less attractive to managers who increasingly opted to use diverse sources. In automobile manufacture, for example, producers exploit a world market for components rather than produce under a single roof. Following the ideas of Rosabeth Moss Kanter (see below), and contrary to the notion of developing long-term stable relationships with suppliers exemplified by Japanese management theory, recent management thinking has emphasized the advantages of flexibility, nimbleness and the minimizing of obligations. Managements have become enthusiastic to outsource work, thereby enabling their organizations to concentrate on key functions – often largely involved with marketing and design (Agar 2004). Extended supply chains have become the norm, not just for the obvious candidates in clothing and footwear, but predominantly in high-tech companies – not least Apple.

Apple Inc. combines the production of highly desirable leading-edge products with exciting designs. Although the company has attracted criticism in recent times as a result of labour problems at its Foxconn suppliers in China, it nevertheless enjoys considerable public goodwill and brand loyalty, its products transcending their immediate purpose and becoming desirable artefacts in their own right. Apple shops have a sort of shrine-like quality and their product launches enjoy the publicity and fervour once reserved for new pop albums. Originally founded in 1976 and operating for 30 years as Apple Computer, Inc. the company removed the word Computer from its name in 2007, becoming just Apple Inc. to reflect a shift in emphasis from personal computers to the wider sphere of consumer electronics. The company is inevitably associated with the charismatic leadership and vision of its joint founder the late Steve Jobs and a series of game-changing products introduced during his era. The company currently produces the Macintosh computers; the iPod, iPhone and iPad as well as a range of software. The emergence of Apple as a market leader in consumer electronics, rather than a niche supplier of personal computers, dates from 1997 – the year Steve Jobs returned to the company following a 12-year absence. In August 1998 the company introduced a new all-in-one computer, the iMac. The design team for the iMac was led by the young British designer, Jonathan Ive, who went on to design the iPod and the iPhone. The iMac, with its translucent plastic casing, was an immediate success and Apple set about aquiring software companies and enhancing its own capacities for software development – resulting in the release of Mac OS X in March 2001. Meanwhile, the company was building an ever-more powerful presence in the marketplace – opening the first of its 350 plus retail stores in May 2001; introducing the iPod in October of the same year (which clocked up sales of 100 million over the following six years); and launching the iTunes store in 2003. iTunes rapidly became the market leader, revolutionizing the music industry and registering five billion downloads by 2008. The iPod has gone through various iterations since its introduction – sales having reached 220 million units by 2009.

The annual Macworld Conference and Expo provides the venue for each Apple product launch. These huge publicity operations centred around the apperance of Steve Jobs wearing his signature black turtle neck and distressed blue denims and were carefully choreographed by Apple's marketing director Phil Schiller. At the 2007 event Jobs announced the launch of the iPhone – a convergence of the iPod and an Internet-enabled smartphone claiming that 'the iPod in 2001 changed everything about music, and now we're going to do it again with the iPhone'. In 2010 Apple introduced their tablet, the iPad, with further iterations launched in 2011 and 2012. In January 2011 Steve Jobs announced that he was taking an indefinite leave of absence from the company owing to ill health and that Chief Operating Officer, Tim Cook, would take over the day-to-day running of the company. Steve Jobs died in October 2011 having led Apple to the point where it was briefly the most valuable company in the world.

THE IMPACT OF THE INTERNET

The Internet has its origins in the Cold War. In the 1960s the US Defense Department's Advanced Project Agency (ARPA) established a series of computer networks for co-operative R&D with the aim of facilitating networking among scientists and government agencies. Initially the Internet, as it came to be called, was hosted by just four computers but the number expanded rapidly to reach approximately 2,000 by the mid-1980s. By the 1980s the administration of the Internet had been taken over by the USA's National Science Foundation (NSF) – the government agency concerned with funding R&D. The NSF initially prohibited commercial use of the Internet but, by the end of the 1980s, growing pressure to allow commercial use had become irresistible and the rules were changed. The NSF withdrew in 1995 and the Internet was effectively privatized – the number of linked networks having grown from around 25 in the early 1980s to 44,000 in 1995. From then on use of the Internet mushroomed, stimulated by the development of the World Wide Web.

The World Wide Web was named by one of its developers, the physicist and computer scientist Tim Berners-Lee. During the late 1980s and early 1990s, Berners-Lee and a team of scientists working at the European Organisation for Nuclear Research (CERN) laboratories in Geneva envisioned a pool of human knowledge which could be stored on the Internet. The pool would be made accessible through a system of hypertext mark-up language (HTML) which could transcend different data formats on various types of computers. The system, through a hypertext transfer protocol (HTTP) would also manage movement of data between Web servers (information storage centres) and Web browsers (McCraw

2000). Similar sites to the one at CERN were set up in the USA, including one at the National Centre for Supercomputing Applications at the University of Illinois. In 1993 a team there, which included the undergraduate student Marc Andreessen, developed an enhanced Web browser which could run on most PCs and also accommodate colour images – they called it 'Mosaic'. In 1994 the 22-year-old Andreessen left Illinois, went into private business and developed a much improved version of Mosaic which was released as Netscape Navigator. Netscape was an immediate success and, together with other browsers such as Internet Explorer, made the Internet accessible to millions of ordinary users as well as the original target audience of engineers and scientists. The arrival of the Internet was a massive game-changer for many companies as e-commerce took off and patterns of commerce shifted from the customary world of bricks and mortar to what the novelist William Gibson described as cyberspace. A potent example is shopping, where consumers have increasingly abandoned conventional shops for suppliers such as Amazon who have successfully harnessed new technology to build a huge and highly profitable business.

ROSABETH MOSS KANTER AND THE MANAGERIAL IMPACTS OF THE INTERNET

In *Evolve: Succeeding in the Digital Culture of Tomorrow* (2001) Kanter attempted to make sense of the changes being wrought by the Internet. In doing so she revisited the themes of innovation and change which had formed the major preoccupation of her work for the previous twenty years or so. Her book *When Giants Learn to Dance* (1989) had outlined the challenges facing organizations in the 1990s and argued that they would have to learn how to achieve more with less resources. The corporations of the future would be far less monolithic and, above all, be characterized by flexibility. Kanter labelled the new approach post-entrepreneurial management and claimed it was based on the following core principles: minimize obligations and maximize options; keep fixed costs low and use variable means to achieve goals; derive power through influence and alliances rather than through full control or total ownership; and keep things moving by encouraging continuous regrouping of people, functions and products to produce innovative combinations.

Evolve addressed the challenges posed by the Internet, a world where Kanter estimated that there would soon be only three types of company which she described as dotcoms, dotcom-enablers, and wannadots. Dotcoms were the pure Internet companies, operating online businesses and existing primarily in cyberspace; dotcom-enablers were the technology providers involved in facilitating and servicing the system; and wannadots were the rest – the existing organizations contemplating or recently embarked upon the great Web adventure. Although at the time Kanter was writing the full impact of the Internet was as yet unknown, there was no doubting the profound effect it was already having on businesses and organizations in general. Kanter was fully convinced that to live with e-culture, as she called it, was to live with change and would require business leaders to possess what she termed seven classic skills: tuning in to the environment, i.e. anticipating the need for change as well as leading it; kaleidoscopic thinking, i.e. stimulating breakthrough ideas; communicating an inspiring vision; coalition-building; nurturing a working team; persistence through difficulties; and spreading credit and recognition. According to Kanter these are more than discrete skills, they also reflect a perspective and a style which is basic to e-culture (Kanter 2001).

CONCLUDING REMARKS

The rapid development of information technology and the intrusion of computers into so many areas of activity changed attitudes and expectations concerning products, production and organizations. Product life cycles became shorter and the arrival of business winners and the departure of business losers ever more rapid. In automotive engineering, improvements in quality were associated with technical and organizational advances which eroded the extreme division of labour which had characterized the industry since the era of Henry Ford. The rise of containerization revolutionized freight handling, slashing costs and massively bosting efficiency. Again, customary patterns of work were swept away in a matter of a few years and once-dominant trade unions reduced to a shadow of their former selves. Containerization removed a long existing bottleneck in trading relationships, massively reducing friction and facilitating the movement of goods globally. These technical changes were matched by a sentiment in favour of free trade and the lowering of tariff barriers, manifested in the establishment of the World Trade Organization in 1995. At the time of writing the free trade sentiment is under pressure and a tendency to protectionism emerging – particularly at the level of political rhetoric. There were of course winners and losers in all of this – literally hundreds of millions lifted out of poverty as a result of industrialization in China but tens of thousands rendered unemployed in the industrialized West where labour costs remained relatively high. Finally, the development of the Internet proved to be the major game-changer of recent times, impinging on pretty well every aspect of life and facilitating whole new patterns of commercial and social activity.

CHAPTER 7

MARKETING POLICY AND STRATEGY

INTRODUCTION

This chapter examines some key elements of business policy and strategy and links them to innovation and marketing. Strategy has become an essential element in the operation of organizations. As the operating environment becomes increasingly complex and volatile, businesses need to reappraise their current situation and attempt in some way to predict future developments. Marketing has gained in significance as consumers have become more sophisticated, the range of products more diverse and the market ever-more competitive. As we shall see in the next chapter there is plenty of evidence to support the view that marketing is involved in a conspiracy to persuade consumers to desire and buy products they do not really need and that advertising is often inaccurate if not deceitful (Packard 1957, Klein 2010). However, there is also the opposing view which holds that in the long term it is impossible to sell consumers products or services they do not want and that marketing is primarily concerned with meeting consumers' requirements (Shaw 2011).

Business policy, strategy, product innovation and marketing all developed in response to ever-greater business and market complexity. As the scale and scope of business activity grew, so the formulation of business policy emerged as a discrete activity and management thinkers began to research, theorize and attempt to generalize about the subject. The need for constant innovation became accepted and the simple process of selling underwent substantial refinement and morphed into marketing. Innovation and marketing will be examined in later sections of this chapter. First, however, we turn to the topic of strategy formulation.

FROM BUSINESS POLICY TO STRATEGIC MANAGEMENT

The basic conceptual framework for the business policy/strategy field was developed by Peter Drucker in his book *The Practice of Management* published in 1954. For Drucker the key management decisions are strategic and centrally concerned with identifying objectives and obtaining the means to achieve them. In a sense this is the difference between management and administration – moving, changing, achieving rather than merely maintaining. Following Drucker's work, and reflecting the military influence and concerns of the time, management writers began to apply the term strategy to what had previously been termed business policy. Alfred Chandler (1918–2007), for example, in his classic text *Strategy and Structure: Chapters in the History of the American Industrial Enterprise*, published in 1962, examined the impact of strategy on organizational structure. Chandler studied almost 100 American firms, tracing the development of these organizations from the early twentieth century to the late 1950s. From this survey Chandler produced extensive case histories of major companies such as Du Pont, Sears, General Motors and Standard Oil. He concluded that changes in corporate strategy preceded and precipitated changes in an organization's structure. In Chandler's classic formulation – *structure follows strategy.*

As with the bulk of management studies, the literature on strategy derived from research in the USA. The pioneer of the subject was Igor Ansoff (1918–2002) who set out the basics in his book *Corporate Strategy* published in 1987. Ansoff was employed by the RAND Corporation where one of his projects involved assessing the vulnerability of the US Air Force's Strategic Air Command to attack by the USSR. After assisting in developing a strategy to cope with that eventuality he joined the corporate planning department of Lockheed Aircraft where he worked on diversification policy. Ansoff subsequently encountered the writings of Drucker and Chandler and developed his notion of corporate strategy, which involved formulating objectives and strategies based on opportunities in the environment. Ansoff developed *gap analysis* (i.e. the gap between where the organization is now and where it wants to be) and popularized the concept of *synergy*, which he described as the 2 + 2 = 5 effect. He also worked on the idea of *stakeholding* which was to become fashionable in the sphere of business ethics during the 1990s (Ansoff 1987). In the wake of Ansoff's work a vogue for organizational analysis emerged rapidly on both sides of the Atlantic. The analytical approach was later criticised by Henry Mintzberg as likely to lead to paralysis by analysis – the making of strategic plans which are never implemented. Whereas Ansoff and his followers favoured a formal, analytical approach to strategy formulation, Mintzberg strongly advocated an informal, intuitive, seat-of-the-pants approach to the subject. Kenichi Ohmae offered a middle way between these two extremes, providing an account of corporate strategy inspired by Japanese experience in his book *The Mind of the Strategist: the Art of Japanese Business* published in 1982. He identified what he termed the *strategic three C's* and the *strategic triangle* – basically the notion that in constructing a business strategy the three main players must always be considered; the corporation, the customer, and the competition; collectively the strategic triangle. Viewed in the context of the strategic triangle, the task of the strategist is to achieve superior performance relative to the competition, in the key factors for success of the business. At the same time, the strategist must be sure that the strategy properly matches the strengths of the corporation with the needs of a clearly defined market. In a later edition of his book Ansoff acknowledged the shortcomings of his earlier work, admitting that he had overestimated the significance of formal analysis at the expense of such issues as leadership, power structures and what he termed organizational dynamics or culture (Ansoff 1987).

CORPORATE STRATEGY – BOEING VS. AIRBUS

In the late 1960s Boeing assembled a group of workers they called 'The Incredibles' – engineers, mechanics, managers and administrative staff tasked with building the largest aircraft in the world – the 747 Jumbo Jet. They completed the task in just 16 months. This immense effort was prompted by a general reduction in air fares, rapid growth in passenger numbers and increasingly crowded skies. The design philosophy behind the new aircraft was (with the exception of the engines) to develop a completely new plane. The 747's final design was provided in three models – all passenger; all cargo; and a convertible passenger/freight model. The massive size of the new aircraft posed many problems. For example, an assembly facility was constructed at Everett, Washington State – 25 miles north of Boeing's headquarters in Seattle. This building, where 747s, 767s, 777s and the new 787 are assembled, remains the largest building by volume in the world – 116.5 million cubic feet. The fuselage of the original 747 was 225 feet long, the tail as tall as a six-storey building and the total wing area as large as a basketball court. The main building was divided into three huge bays, each 1,000 feet long, 300 feet wide and 115 feet high. One bay housed the sub-assembly areas for the wings and sections of the fuselage, while the other two bays each housed a final assembly line with stations for four complete aircraft. The bays were linked and served by 11 30-ton capacity overhead cranes used to transfer the sub-assemblies through various stages of build-up and eventually carry the completed sub-assemblies to the start of one of the final assembly lines (Lucas 1988). Much of the work on the 747 was subcontracted out to various suppliers – in all there were dozens of major subcontractors, 1500 primary and 15,000 secondary contractors. In total a 747 contains 4.5 million parts and production time for each aircraft is approximately 20 months, only the last two of which are spent on the final assembly work. Much of the assembly work is completed at the sub-assembly stage, so that the major sub-assemblies are more or less complete by the time they are brought together on the final assembly line.

The 747 outflanked its more glamorous contemporary competitor, the Anglo-French Concorde, and remains in service today. However, by the 1990s a debate had begun regarding the possibility of producing something even bigger – a 'Super Jumbo'. In 1993 Boeing and Airbus jointly agreed to commission a feasibility study to estimate the market for such an aircraft and review the likely development costs. Interestingly the two companies interpreted the outcome of the joint study entirely differently. Whereas Airbus predicted a possible market for 500 to 600 planes,

Boeing envisaged likely sales of only 300 to 350. Both companies agreed the likely development costs (at 1992 prices) would be in the region of US$14.5 to 15 billion with negative cash flow reaching US$22 billion (Newhouse 2008). Briefly, this meant that the aircraft would be sold at a massive loss for some unknown period before the development costs could (if ever) be recouped. Whereas Airbus were persuaded to take the plunge Boeing were hesitant, some executives doubting whether there was a market sufficiently large enough to justify the initial cost of developing the new aircraft and others feeling that the company already had enough on its hands with the need to modernize the 737 and develop the 777. In the event, lacking a clear strategy, Boeing decided to play for time talking up various new developments for the 747 and ultimately opting to build the 787 Dreamliner, incorporating new light, composite materials and higher performance engines. Whatever the internal issues relating to strategy, the decisions to build the two aircraft demonstrated two differing visions of the market for airliners.

The A380 was orientated towards the emerging market in the Far East – flying large numbers of passengers to hub airports (the hub and spoke theory) – many of which have to be enlarged to accommodate it. The A380 made its maiden flight in 2005 and went into service in 2007. At the time of writing around 100 have been delivered – priced at approximately US$390 million each – and it is estimated that Airbus will need to sell 420 units to break even on the costs that have so far accrued. The Boeing 787 was calculated to be energy efficient, comparatively environmentally friendly (a gesture in the direction of the environmental lobby) and to fly direct to the destination airport – the point to point theory. The 787 made its maiden flight in 2009 and entered service in 2011 – at the time of writing 50 have been delivered. A 787 costs approximately US$200 million. Both aircraft have suffered delays and technical problems – the A380 has suffered cracks in the wings while the 787 was grounded as a result of problems with the plane's lithium-ion batteries. Meanwhile, Airbus has developed a direct competitor for the 787 in the shape of the A350, which at the time of writing has just completed its maiden test flight.

MARKETING STRATEGY – PRODUCT, PRICE, PROMOTION, PLACE

A useful way of approaching marketing at the strategic level is via the '4Ps' – Product, Price, Promotion, Place. Analysis of each of these elements leads on to consideration of the following interlinked factors (Shaw 2011): the customer (a successful marketing strategy requires as much information as possible about current and potential customers, including market size, demographics, customer expectations and attitudes) and product design and development – these two when linked to an active, innovative and relevant R&D programme, will feed into marketing strategy, part of marketing's role being to work the market for the introduction of new products; pricing and revenue management – the crucial issues of income and cash flow are self-evidently linked to the size of the marketing budget, while pricing and the appropriate point of entry into the market are also key issues for consideration; distribution channel selection and control once the overall strategy has been decided decisions on how to distribute the product or service can be made. Issues such as customer service can be considered in relation to marketing, selling, advertising and promotional policies.

Globalization and the Internet have widened consumer choice and stimulated the possibility of shopping around for products and services – an example being price comparison sites. Philip Kotler (2000), the leading marketing guru, has identified the major elements or practices which can contribute to developing a successful marketing strategy as follows:

- *Higher quality* – poor quality is always bad for business and a reputation for poor quality lasts indefinitely. For example, General Motors is still suffering from having a reputation for bad quality long after the problems have been solved. In contrast, Toyota has managed to sustain a reputation for producing high-quality vehicles even though they have experienced technical problems and several product recalls in recent years. A further issue relates to whether customers are willing to pay the price necessary to sustain the highest quality, which equally depends on where the product is pitched in the market:
- *Better service* – everyone wants good service but perceptions vary between customers. For example, customers in Burger King will have a different perception of good service than, say, the customers in Claridges. Equally, passengers on budget airlines have different expectations from those flying business class with a major airline. There is always a risk in emphasizing service quality and failing to deliver:
- *Lower prices* – lower prices are an obvious source of competitive advantage. Wal-Mart, still the world's largest retailer, uses its buying strength to sustain low costs. Retailers such as Primark and T.K. Maxx offer no frills stores which emphasize low costs rather than the enhanced shopping experience provided by Harrods or Fortnum and Mason. Price, however, is a very sensitive issue and brands can easily be damaged by being associated with cheapness:
- *Large market share* – gaining substantial market share can obviously be advantageous. In general terms market leaders make more money than their competitors and enjoy greater brand recognition. However, large market share does not always sustain commercial success – see as examples IBM and General Motors:
- *Adaptation and customization* – customers often want suppliers to modify products or services to meet special requirements. Hotel chains such as Holiday Inn provide a number of adjustments to the standard package

which are calculated to meet individual tastes:

- *Continuous improvement* – we have encountered the idea of continuous improvement in the context of total quality management. As Apple have shown, enhancement of a product or service can help to keep customers interested. Having said this, however, products can reach the point where further improvements are met with indifference from consumers – particularly if the price is hiked. For example, how many blades are really necessary in order to get a decent shave?:
- *Entering high-growth markets* – high growth markets offer glamour and possibly fabulous wealth but the failure rate is also high. A further problem is that products can become obsolete very quickly in fast-growing industries and companies must go on investing in order to keep up. Equally, it is difficult to take the profits from a product before having to invest in the replacement:
- *Exceed customer expectations* – exceeding customer expectations is a reliable, if difficult, strategy. The problem is to go on meeting ever-rising expectations while maintaining profits and prices. Exceeding customer expectations is one of the keys to ensuring returning customers and repeat purchases. A reputation for high levels of customer service is a great advantage to the business and should therefore be considered as a benefit rather than a cost – see for example the case of AMEX which follows.

AMEX AND IMPROVED CUSTOMER SERVICE

Credit cards have a long history reaching back into the early years of the twentieth century (Gross 1996). However, the first modern example was the Diners Club card launched in New York in the 1950s which, for a modest fee, enabled holders to purchase meals at participating restaurants. At that time American Express (AMEX) was primarily a travel company which issued travellers' cheques and they were initially reluctant to become involved in this new departure. Nevertheless, the success of Diners Card and the launch of other competing products prompted AMEX to act, launching its first credit card in 1958. The initial results were disappointing and it was not until George Waters (1916–2003) joined the company in 1961 that things began to improve. Waters had worked for IBM in sales and marketing and had also been chief operating officer for a grocery chain. He set about increasing the annual fee to hold a card, cracking down on late settlement of outstanding balances and improving terms for retailers. He also had the card itself redesigned and retained the leading advertising agency Ogilvy, Benson and Mather to develop a marketing strategy which included the tag line 'Don't leave home without it'.

By the end of the 1960s the AMEX card had become the market leader and the company's earnings were growing at 17 per cent per annum (Gross 1996). When Waters eventually retired in 1980 the company honoured him with the title 'father of the American Express Card'. In the early days of American Express the company had emphasized customer service and advertising campaigns were constructed around the safety net provided by American Express traveller's cheques and, later, possession of an AMEX card. Eventually competition eroded all of this to the point where customer service was seen as a necessary evil – something which had to be supplied but not a constructive element in the company's business model. However, following the results of a survey in 2012, the AMEX view of customer service changed entirely. Customer service is now seen as a means of networking with customers and better profiling their needs. The four major service complaints which emerged from the survey were as follows: *rudeness* – basically encountering an insensitive or unresponsive customer service representative; *passing the buck* – being shuffled around the system with no resolution of the issue; *the waiting game* – being kept waiting for an unacceptably long time to have an issue resolved and, finally, *being boomeranged* – being forced to continually follow up on an issue.

Apparently the average time Americans are willing to wait for customer service before slamming down the telephone is estimated at 13 minutes. Similarly, they will wait for an estimated 12 minutes for in-person help in banks, restaurants and shops before giving up. Loss of a customer at this point is depicted by shopping guru Paco Underhill as one of the worst types of failure in the retail world – having got the customers into the shop and persuaded them to buy, only to see them abandon the purchase due to the wait at the checkout being futile (Underhill 1999). AMEX offers the following suggestions for a company wanting to offer better customer service and improve how it is viewed by customers: great service starts with the people who deliver it; it's all about relationships; make it easy for customers to do business with you; exceeding expectations is easier than you think; listen to your employees; seek opportunities to make an impression.

MARKETING AND THE DIFFUSION OF INNOVATION

The final element in Kotler's list of elements is perhaps the most important – product innovation. As we have seen, a central characteristic of the modern world is the constant pace of innovation and we have encountered numerous examples throughout this book, ranging from the emergence of the factory system to the widespread adoption of containerization and from the development of the moving assembly track to the application of robotics. Innovations occur in certain places at certain times and spread to other areas either gradually or more rapidly depending on circumstances. This process is referred to as *diffusion*. Like individuals some societies are more comfortable with innovation than others and this fact can inhibit or accelerate the pace of diffusion. In his book *Diffusion of Innovations* Everett Rogers (2003) identifies the four main elements in the diffusion of innovation as follows:

- *Innovation* – the nature of the innovation itself will influence the pace of diffusion. Innovations which have an apparent relative advantage over what is currently on offer, are compatible with existing values, past experiences and present needs, are simple to understand and can be tried with minimal risk have an advantage in terms of diffusion. An example is the QWERTY keyboard developed in the early 1870s by Christopher Latham Sholes (1819–90) with the purpose of reducing the jamming of typewriter keys but which was subsequently accepted as the international standard and, finally, adopted with minor modifications for computer keyboards.
- *Communication channels* – communication is the process by which participants create and share information with one another in order to reach a mutual understanding. A communication channel is the means by which messages move from one individual to another. Communication channels can be through mass media or interpersonal channels. In any case shared language will facilitate diffusion.
- *Time* – refers to the rate of adoption of any innovation in any particular set of social circumstances. This incorporates the innovation-decision process which runs from initial knowledge of the innovation; being persuaded to try it; deciding after a trial period to adopt or reject it; implementing it for a trial period; and confirming its use over time. Rogers coined the term 'early adopters' to describe the individuals or groups who are most willing to try an innovative product or service.
- *A social system* – may be made up of just a few individuals, groups or entire societies. One aspect of a social system is the development of norms of behaviour which can promote or inhibit the diffusion of innovation – birth control being an obvious example.

In the modern world innovation is intimately linked to marketing which seeks to accelerate the process of diffusion while attempting to create a social climate favourable to a new product. We have become comfortable with a world in which we are constantly reminded of brands and assailed by continuous advertising calculated to urge us to either try a new product/service or continue to purchase what we already have (i.e. brand loyalty). The acknowledged pioneers in developing these aspects of marketing were Procter & Gamble.

PROCTER & GAMBLE – BRAND MANAGEMENT AND MARKET RESEARCH

Brands are not a new phenomenon. The design on the can of Tate and Lyle's golden syrup was first used in the 1880s and remains virtually unchanged. In Eduard Manet's (1832–83) 1882 painting *A Bar at the Folies-Bergère*, currently in the Courtauld Gallery in London, the label on a bottle of Bass beer is clearly identifiable. However, it was Procter & Gamble who advanced the notion of brand from something merely calculated to denote a product's authenticity to something far more profound and at the heart of modern retailing. Procter & Gamble is a major multinational corporation producing a wide range products in two major sectors,beauty and grooming and household care. Many of the products are household names such as Daz, Fairy, Bold, and Head & Shoulders. Others are niche products such as Lacoste, Dunhill and BOSS fragrances. P&G's customary approach has been to present the individual brands as the face of the company. However, as a Worldwide Olympic Partner to the 2012 London Olympic Games the company decided to promote the profile of P&G as a whole. P&G has built its portfolio of brands bit by bit over the course of the last century – in 2005, for example, the company acquired Duracell, Gillette and Oral B for a staggering US$57 billion.

P&G have always been very active in advertising, including in the past sponsoring radio dramas (the original soap operas), running various contests, operating door-to-door giveaways of coupons and arranging for thousands of salespeople to visit retailers and promote demand (McCraw 2000). It was as a result of successes with major advertising campaigns that P&G developed a new business technique – *brand management* – which focused attention on the product rather than the business. Brand management had a similar impact to Sloan's multi-divisional organization structure by tending to decentralize decision making. In May 1931 Neil McElroy (1904–72), a young Harvard graduate who had worked on advertising Camay soap, circulated an internal memorandum which argued that each P&G brand should have an individual manager responsible for promoting that specific brand and no other. Each brand should as far as possible be targeted at different consumer markets by deploying what is now termed *product differentiation*. This idea was calculated to avoid cannibalizing the brand – reducing market share, sales volume or sales revenue by setting two brands against each other in the same segment. Having said this, however, P&G have stimulated competition between their brands, most recently between Mach3 and Fusion razors, with a view to expanding the entire market.

McElroy's proposal was adopted by P&G and subsequently emulated by many consumer-products companies around the globe – balancing central oversight with decentralized decision making on the model pioneered by Alfred Sloan (McCraw 2000). Neil McElroy's formula for P&G's success may be summed up as 'find out what the consumers want and give it to them'. This obviously requires accurate information about the market and still characterizes P&G's approach to product development. The leader of the company's market research effort for many years was D. Paul 'Doc' Smelser (he had a Ph.D. in economics from Johns Hopkins University) who built a team of market researchers which grew from a tiny original core to eventually number several hundred. Smelser's staff carried out door-to-door field research asking consumers detailed questions about P&G

products and also monitoring and tabulating the impact of advertising campaigns. In the 1960s P&G began using telephone and postal services to conduct mass surveys and, by the 1970s, the market research department was carrying out over a million and a half interviews annually. As TV advertising grew in importance P&G used the so-called 'DAR (Day After Recall)' method to measure the impact and memorability of its commercials. Self-evidently the activities of McElroy and Smelser at P&G were aimed at persuading consumers to purchase and go on purchasing P&G products, and the techniques they perfected have been widely adopted by numerous other companies.

CONCLUDING REMARKS

The pace of technical change has been very fast in recent years and this is likely to continue. Global competition is set to increase as China and India emerge into world trade and begin to make serious inroads into sophisticated markets. Change is linked to timing. Evolutionary, gradual change can be achieved if time allows. However, increasingly, revolutionary change is occurring whereby things change so quickly that the customary way of life fractures and the result is perplexity and bitterness. In the long term most products and most companies die, but organizations can take action to avoid unnecessary fatalities.

CHAPTER 8

MARKETING AND ITS DISCONTENTS

INTRODUCTION

This chapter examines some of the criticisms of marketing and the so-called throw-away society. An old nostrum of marketing is that it is impossible to get a person to buy something they do not want – or at least to buy it again if they are not satisfied with it. Having said this, marketing has become increasingly sophisticated and brands have become a major means of navigating our way through life. Those who make their living by selling us things would respond to criticism of what they do by claiming that they are only providing useful information. The view that business succeeds best when it closely considers the needs of consumers is now firmly rooted and it appears to be self-evident that an organization must satisfy its customers or customers will merely defect to competitors. However, things are not quite that simple. The ethics of marketing and the developing relationship between producers and consumers were examined during the 1950s and 60s by the muck-raking American pop sociologist Vance Packard (1914–96). Packard was writing in a period when the USA was emerging fully into what is now called consumer society, a period before Britain and much of Europe had recovered from the impact of the second world war. In *The Hidden Persuaders* (1957) Packard studied the development of so-called 'Orwellian' techniques of subliminal messaging and the application of psychological insights to the processes of selling. One of his insights was that what we view as rational choices (and defend as such) are very often just impulse buys.

Interestingly the most memorable passage in *The Hidden Persuaders* is Packard's account of the Philip Morris Company's 'Marlboro Man' advertising campaign which rapidly transformed a ladies' cocktail cigarette, which had been introduced in 1924, into a macho brand associated with cowboys, the outdoor life and ranching. Originally the brand had been calculated to appeal to a new generation of women smokers with posters depicting a sophisticated flapper, looking rather like a character from Baz Luhrmann's recent film *The Great Gatsby*, sporting a fashionable cloche hat and cigarette holder and deploying the tag line 'Mild as May'. The theme of the woman of fashion was continued when filtered tips (reflecting a growing awareness of the health risks associated with smoking) were introduced and using the tag line 'Ivory Tips protect the lips'. By the post-Second World War period the brand had been repositioned to appeal to a new generation of mothers with pictures of babies and captions such as 'Before you scold me, Mom ... maybe you'd better light up a Marlboro' and 'Gee, Mommy you sure enjoy your Marlboro', accompanied by a smiling mother and, given what we now know regarding the hazards of smoking, the chilling tag line 'Yes, you need never feel over-smoked ... that's the Miracle of Marlboro!'. The campaign which eventually created 'Marlboro Man' began in 1954 and was master-minded by Leo Burnett (1891–1971). Burnett was inspired by an article in *Life* magazine which told the story, accompanied by photographs taken Leonard McCombe, of a Texas cowboy, Clarence Hailey Long (1910–78). Long bore a close resemblance to the actor John Wayne (1907–1979) whose performance in John Ford's (1894–1973) 1939 film *Stagecoach* embodied many of the attributes of freedom, masculinity and toughness combined with a kind of taciturn charm which Burnett wanted to associate with the new Marlboro. Further, the campaign coincided with the immense popularity of TV Westerns such as *Gunsmoke*, *Wagon Train* and *Cheyenne* all calculated to appeal to the yearnings of an increasingly urbanized and sedentary population and often accompanied by a stirring theme tune sung by Frankie Laine (1913–2007). In the event the Marlboro Man campaign, with its associated tag line 'Come to where the flavour is. Come to Marlboro Country', proved to be one of the most successful and influential in advertising history and ran for 45 years. In 1955 sales of Marlboro were US$5 billion – by 1957 they had risen to US$20 billion. Further, the success of the campaign enabled Philip Morris, and other tobacco companies which followed their lead by devising sexy advertising campaigns, to overcome for a while the growing health concerns connected with cigarette smoking. In a sense Vance Packard was correct regarding the power of advertising to persuade consumers. However, there was nothing hidden about the Marlboro Man campaign which, although originally conceived as appealing to men, also attracted women and established Marlboro as one of the most recognizable brands on the planet.

THORSTEIN VEBLEN AND THE THEORY OF THE LEISURE CLASS

Thorstein Veblen (1857–1929) published *The Theory of the Leisure Class*, his critique of the USA's emerging consumer society, in 1899. Partly his target was the new class of wealthy businessmen which had emerged in the post-Civil War USA – the so-called robber barons who were ruthless, acquisitive and more interested in making money than producing reliable goods or services. For Veblen such men were the inheritors of a long tradition of groups, castes and classes which had risen to dominate their societies. As part of his analysis Veblen put forward three characteristics of such dominant groups – *conspicuous leisure*, *conspicuous consumption* and *conspicuous waste*. Display and ritual were central to all of this, providing opportunities to demonstrate apparent material (i.e. pecuniary) superiority by a show of expensive possessions and lavish entertainment. Clothing performs a key function in such display as does jewellery and even perfume – all indicators of a form of superiority. The result of this, according to Veblen and contra Marx, is not to generate resentment among the 'lower orders' but to generate acceptance, approval and the desire or aspiration to emulate their apparent 'betters'.

Veblen's views were driven by a sort of tolerant contempt for the games people play. He was an outsider and, in his later writings, argued that the prevailing social system would eventually be swept away by technological change and the rise of a rational order order led by technicians and engineers. Further, *The Theory of the Leisure Class* contains within its satire a moral view, also shared by Packard, based on a preference for a simpler way of life freed from the many vanities and falsehoods which characterized contemporary America. Veblen was writing at a time when a debate was beginning in the USA on the issue of overproduction. The emerging techniques of mass production were producing more goods than could readily be consumed. We hear this problem echoed today in calls for economic growth linked to stimulating demand. One answer for individual producers was to generate demand through advertising, branding and the creation of disposable products, which leads us to the work of King Gillette (1855–1932).

GILLETE AND THE DISPOSABLE RAZOR BLADE

By differentiating the product through building a recognizable brand producers can persuade consumers to repeat purchase a product, i.e. encourage repetitive consumption. Obvious candidates for this approach are cigarettes, alcohol and patent medicines. However, for the approach to succeed, the product has in some way to be differentiated from those of competitors – hopefully building brand loyalty. Often the differentiation is spurious, consumers' preference stemming not from the product itself but the nature of the advertising and brand image – how many people can really tell one cold lager from another? Maintaining demand and growing the customer base encourages producers to develop a strategic approach – including the strategic decision to produce products calculated to become obsolete. At the time when modern brands were emerging in the early twentieth century, producers were already developing disposable products calculated to guarantee repetitive consumption. An obvious example is the disposable razor blade, which, in spite of the availability of electric razors, still commands a massive market – it is estmated that two billion blades are used each year in the USA alone.

The Star safety razor, introduced in the late nineteenth century, was certainly safer than the traditional open, cut-throat razor but it still had the disadvantage of needing to be professionally re-sharpened. In 1895, King Gillette hit upon the idea of producing a cheap, thin blade which could be stamped out of metal and then honed on two edges. However, realizing this idea posed considerable engineering problems which took six years to solve. Eventually a process was developed which interleaved sheet steel and copper, allowing the thin metal sheets to temper without buckling. Only tempered sheets could hold the razor-sharp edge required for Gillette's disposable blades. In 1905, by which time the disposable blade was well on the way to general acceptance, Gillette launched the tag line 'No Stropping. No Honing'. In a sense this was a shaving revolution and Gillette opened factories in Canada, Britain, France and Germany to meet the global demand for his blades. He also adopted the 'razors and blades' business model whereby razors are sold cheaply in order to increase the market for blades. Like other disposable products – paper handkerchiefs for example – Gillette's blades were convenient and also hygienic. The development of disposable products brought many positive advantages – as well as encouraging a throw-away mentality and generating huge quantities of waste.

PLANNED OBSOLESCENCE – TECHNICAL AND PSYCHOLOGICAL: EXAMPLE GM VS. FORD

The emergence of planned obsolescence was more dubious in terms of the balance of advantage between producers, consumers and the wider society. A significant development was the introduction of the annual model change by General Motors during the 1920s. Whereas Henry Ford, notwithstanding his riches, encouraged thrift and built cars to last, the head of GM, Alfred Sloan, saw things somewhat differently. Sloan subscribed to the notion of technological obsolescence, the view that new products should inevitably replace the old and that one of the main purposes of R&D was to promote this process – even if a product survived it would need to be new and improved if it were to maintain market share. A key example from the early years of the automotive industry is the electric starter, patented by Charles F. Kettering (1876–1958) in 1913, which rendered hand-cranked cars obsolete more or less overnight. The innovation also opened a new market among potential women drivers who were unable (or unwilling) to crank start a car. Kettering became head of R&D at GM in 1920, a post he held until 1947. When Sloan became president of GM in 1923 he and Kettering set about rendering Ford's Model T obsolete. Kettering had made a failed attempt to achieve this with the 1921 Chevrolet but the vehicle was a technical failure. For the 1923 buying season, however, GM engineers totally repackaged the car, giving it some of the stylistic lines of an expensive vehicle of the era. The new Chevrolet was a huge success and rapidly rendered the Model T a thing of the past. Sloan became convinced that technological obsolescence was only one (and perhaps the most expensive) way to market new cars. A far cheaper and equally effective approach was to change an automobile's styling, thereby rendering the vehicle out of date immediately – what is known as psychological obsolescence (Slade 2006). Whereas Ford's 'Rouge' plant was designed to produce a single vehicle,

GM adopted a flexible mass production system far better equipped to respond to continual model changes.

Psychological obsolescence based on styling, rather than technical superiority, became central to GMs continuing success and Harley J. Earl (1893–1969) was crucial to this approach. Earl joined the company in 1925 having studied engineering at Stamford University and worked in his father's coach-building business as a designer of customized cars for Hollywood celebrities. His early work for GM included the Cadillac La Salle, the first mass-produced car designed by a stylist. Earl became the inaugural head of GM's Art and Colour Section, which later became the Styling Section – growing from an initial staff of 50 to reach more than 1,000 by the time of Earl's retirement in 1959. Earl was closely linked to the annual model change policy, using such styling features as massive chrome radiator grilles, sweeping tail fins and brake light clusters – all excessive and symbolic of the USA's post-Second World War affluence. Psychological obsolescence is at least partially driven by fashion. Today's must have is tomorrow's candidate for recycling – until, that is, it becomes vintage. Psychological obsolescence was a central element in youth culture as it developed in the USA in the 1950s and spread around the world in the decades which followed. The discovery of teenagers as a distinct market with money to spend stimulated a massive growth in products specifically aimed at that group. Specific films, clothes, cosmetics, drinks, motorcycles and automobiles were all specifically designed and marketed to teenagers. Above all, however, it was music which provided, quite literally, the unifying theme. The widespread availability of cheap record players for teenagers' bedrooms, together with juke boxes for diners and coffee bars, made music the essential ingredient in establishing teenage identity. As we saw in Chapter 3 the single most significant point in this development was the release by RCA of Elvis Presley's *first* album in March 1956. Its massive success opened the door to numerous Elvis clones on both sides of the Atlantic. Rock music was calculated to be ephemeral – to be played literally to death in a matter of weeks and then replaced by the next big thing – of course nothing had changed, the song was still the same but, as Eminem puts it, it had become 'cold product' – a classic case of psychological obsolescence.

INSIGHTS FROM MARCUSE AND BRAVERMAN

Marketing, like the economic system which created it, has always attracted critics from the left of the political spectrum, two of the most potent being Herbert Marcuse (1898–1979) and Henry Braverman (1920–76), both of whom were inspired by Karl Marx (1818–83). Marcuse was associated with the emergence of the so-called New Left in the USA during the 1960s. In part this movement was stimulated by the USA's involvement in the Vietnam War. Partly too it sprang from domestic considerations, including civil rights issues and feminism. The then USSR, the motherland of the Old Left, was seen as being tainted by the ravages of Stalinism and, in the aftermath of the 1956 suppression of the protest movements in Hungary, the type of imperialism previously pursued by some of the European nations in Africa and Asia and currently by the USA in Vietnam. Many of the New Left's attitudes to contemporary capitalism were summed up in Marcuse's best known book, *One-Dimensional Man: Studies in the Ideology of Advanced Industrial Society*, published in 1964. Marcuse was frankly pessimistic about the possibility of a revolutionary transformation of society ever occurring in the industrialized West. The success of Western capitalism, he argued, had created a situation where the vast bulk of the population were materially satisfied and seduced by the possession of consumer goods. Although the classic Marxist dichotomy between the bourgeoisie and the proletariat still existed it had been stripped of revolutionary potential. In Marcuse's view social unrest would in future come from those on the margins of society, the outsiders for whom Marx had allocated no historical role. However, although the populations of the industrial societies might be better off than ever before they were far from being fulfilled. On the contrary they were estranged from their work, each other and ultimately themselves – in a word they were alienated. Further, according to Marcuse, they were trapped in what he described as a totalitarian social system which avoided naked political terror but deployed techniques of manipulation instead. Basically they were caught in a system where the production and consumption of waste was paramount and prevailed over the human desire for fredom – or as the humanistic psychologist Abraham Maslow (1908–70), Marcuse's sometime colleague at Brandeis University put it, 'self actualization' (Sheldrake 2003).

Marcuse's philosophy helped create an intellectual climate sympathetic to Braverman's account of alienation through the degradation of work in his *Labour and Monopoly Capital: the Degradation of Work in the Twentieth Century* published in 1974. Braverman began his analysis of what he saw as the degradation of work (basically an ongoing process of de-skilling) by briefly tracing the origins of wage labour. He argued that workers are inevitably forced into paid employment because prevailing social conditions leave them with no other way of making a livelihood. Meanwhile the employer, as the possessor of capital, pursues profits and, in order to obtain them, seeks to dominate the labour process. Whereas capitalists can readily evaluate the value of buildings, machinery, tools and materials in the labour process, they are unable to apply similar precision to the utilization of their workforce. This is because capitalists do not purchase workers as such but rather only the workers' time. In order for the capitalists to gain profit from their activities it is obviously essential that the workers produce more than they consume in wages. However, the amount produced is susceptible to variation depending upon the extent to which the workers apply themselves to their tasks. This subjective element is just as likely to lead to restriction as it is to maximization of output. Hence capitalists are forced to seek means of minimizing workers' control over their work by developing techniques to increase certainty of performance. Braverman readily conceded that employers had always sought to control their workers. However, he believed that this search had reached its highest level with the development of scientific management. Automation only serves to reduce the level of workers' skill even further. Inevitably, technological innovation under capitalism will be increasingly applied as a means to increase control over workers and also as a means of generating profit. Alongside the ever greater control over the labour process exercised by capital, Braverman claimed that a universal market had emerged and society was being transformed into a gigantic marketplace where everything becomes a commodity. The universal market and the degradation of labour become complementary phenomena, ensuring that the workers generate profit at both the point of manufacture and consumption. Braverman does not offer any escape from the situation he describes through a revolutionary transformation of society but merely concedes capital's absolute domination over labour (Sheldrake 2003). Whereas Marcuse saw marketing as a key part of the process whereby the inhabitants of advanced capitalist societies were more or less anaesthetized by consumerism, Braverman emphasized the limited control individuals had over their lives as a result of declining levels of competence and an artificially induced appetitie for commodities. In the period since Marcuse and Braverman were writing there has been a shift of much of the so-called degraded work to cheaper labour markets, leaving behind a constant problem of unemployment but thus far no reduction in the appetite for consumption. Ironically, China is the leading force in this shift and remains, nominally at least, a Communist country.

CONCLUDING REMARKS

Marketing has attracted tremendous criticism, some of the main points of attack being that it is exploitative in numerous ways, plays on people's vulnerabilities, perpetuates an inherently unequal social system and encourages debt. The associated world of advertising is also criticised – although in Britain the industry is regulated by an Independent Advertising Standards Authority charged with ensuring that advertising is legal, decent, honest and truthful. Since 2011 product placement, a regular element in the funding of films, has been permitted in British TV programmes. Again, this has been criticised as a further intrusion of commercialization. Of course, marketing must shift product or producers would not spend such vast amounts on advertising – for example, each time Apple launches a new phone it expects to sell as many units as all of the previous versions combined, so perhaps it's not surprising that its advertising budget increases accordingly. The company's spend on advertising was US$501 million in 2009, rising to US$691 million in 2010 and US$933 million in 2011. So marketing and associated advertising are very big business. However, the extent to which ever-more savvy consumers are duped by advertising and marketing generally or are complicit in the process – knowing that they are being prompted to buy things for a variety of reasons other than utility – remains open to debate.

CHAPTER 9

SUSTAINABILITY

INTRODUCTION

This chapter examines some of the environmental and sustainability issues associated with science and technology. Concern over the environment and the impact of urbanization and industrialization is not a new phenomenon. Pollution was once accepted as an inevitable and almost honourable concomitant of industrial success. During the long process of industrialization, urbanization and population growth in Britain from the late-eighteenth through the nineteenth centuries, there was no lack of literary comment on the filthy, degraded world which seemed to be emerging – associated with what the poet William Blake (1757–1827) described as the 'dark satanic mills' or factories. Much of this material was overtly nostalgic and seemingly yearning for a lost rural idyll which almost certainly never existed. Nevertheless, by the mid-nineteenth century the impact of the badly polluted water supply in London (not least epidemic diseases, most notably cholera) was acknowledged in the construction of the vast network of sewers still in operation today.

During the nineteenth and early twentieth centuries a huge effort was made to clean up London, shifting everything from people, garbage and even the dead out of the central core of the city to areas beyond its boundaries – essentially down river to the east. Although the health risks associated with a badly polluted atmosphere were known of as early as the 1930s, when a campaign had been mounted to resist the building of Battersea Power Station in inner London, it was not until the 1950s that they succeeded in forcing their way on to the political agenda. A series of 'smogs' occurred, culminating in the Great Smog of 1952 which is estimated to have killed 4,000 people in the short term and a further 8,000 in the immediate period which followed. Initially, the British government blamed the deaths on an influenza epidemic but the weight of medical evidence eventually prompted the Clean Air Act of 1956 which declared smokeless zones and inaugurated a new era of environmental awareness. The Clean Air Act was the beginning of the end for the British coal industry (Sheldrake 1991) as domestic users converted to natural gas and electricity while commercial users, such as shipping and railways, abandoned coal for oil and electricity. New-build power stations were generally oil-fired and those which burned coal were depicted as environmentally hostile. Also, the first generation of nuclear power stations came on stream with the construction of Britain's first commercial power station at Calder Hall in 1956. Having said all of this environmentalism was hardly an ideology, the dominant mood being a strong belief in the benefits of modernity, technology and science.

THE CHALLENGE OF ENVIRONMENTALISM

A potent challenge to the apparently unassailable advantages of science came in 1962 with the publication in the USA of Rachel Carson's (1907–64) highly controversial book *Silent Spring* which purported to expose the hazards of DDT. Carter was a marine biologist and ecologist and her book claimed to describe how DDT enters the food chain and accumulates in the fatty tissues of animals, including human beings, raising the possibility of cancer and genetic damage. Following the increased awareness generated by Carson's book a rather more measured approach was taken by Kenneth Mellanby (1908–93) in his *Pesticides and Pollution* published in 1967. Mellanby concluded his book on a positive note, claiming that the impact of pollution could be reversed if urgent action were taken. Contemporary with Carson and Mellanby's work was the tragedy of the drug thalidomide, which had been supplied to pregnant women to combat morning sickness. Unfortunately the clinical trials of the drug had not been properly performed. In the event about 15,000 foetuses were damaged, of which about 12,000 in 46 countries were born with birth defects, with only 8,000 surviving past the first year of life. Finally, in this unfortunate cocktail of chemicals, mention should be made of Agent Orange (2,4,5 T) which was used between 1961 and 1971 as a defoliant in connection with the USA's programme of herbicidal warfare during the conflict in Vietnam. Agent Orange proved to possess carcinogenic properties and also to be a potent source of growth defects among people and animals exposed to the chemical. Meanwhile, the ecologist Frank Fraser Darling (1923–79) in his 1969 Reith Lectures, *Wilderness and Plenty*, warned against the growing problems associated with rapid population growth and general environmental degradation leading to the unsustainable loss of habitats and species.

Perhaps not entirely by coincidence the two major environmental campaign groups were formed at around this time, as people in the developed countries began to react against the claims of technology and science to deliver a better world. Friends of the Earth were founded in Britain in 1971 and currently claim to have over 2 million members in 74 countries. Greenpeace was founded in Canada (its headquarters are currently in the Netherlands) in the same year and claims a membership of 2.8 million in 41 countries. In terms of national politics, only the German Green Party has so far made any real impact. Nevertheless electorates and consumers are now highly sensitized to environmental issues, with both of the major parties in Britain having deployed ideas of 'red-green' and 'blue-green' policies in their political rhetoric.

INSIGHTS FROM C. P. SNOW AND FRITZ SCHUMACHER

C. P. Snow (1905–80) is primarily remembered as a novelist, most notably for his 11-volume sequence *Strangers and Brothers* published between 1930 and 1970. However, during the 1960s he became involved in stimulating public debate on the role of science in society and the weakness of the British educational system in addressing the challenges of an increasingly technological society. Although in the 1950s the world of work in Britain remained much as it had been between the wars, new technologies were developing which threatened the old, labour-intensive industries of the past. Applied science was poised to become increasingly important and industrial R&D ever-more crucial to economic success in emerging industries such as computers, atomic energy, telecommunications, electronics and pharmaceuticals. The USA remained pre-eminent in the sphere of industrial R&D, while Germany and Japan had made a remarkable recovery from the destruction of the war. The former in automotive, chemicals and electrical goods, the latter in electronics where Japanese manufacturers were successfully producing vast quantities of cheap transistor radios – by 1959 Japan was exporting six million radios a year to the USA and proportionally as many to Britain. The old order was rapidly changing and it was in this context that Snow gave his famous, notoriously controversial and lastingly influential Rede Lecture *The Two Cultures* in Cambridge in 1959. Snow had begun life as a scientist, attending Leicester's University College where he took a First in chemistry in 1927 and an M.Sc. in physics in 1928, before gaining a scholarship to Cambridge, completing a Ph.D. in Physics in 1930 and becoming a Fellow of Christ's College in the same year. Cambridge and science form the background to several of Snow's books – most prominently *The Masters* (1951) and *The New Men* (1954), the latter dealing with the development of Britain's atomic bomb. His most famous work *The Corridors of Power* was published in 1964 and it can be argued that it was during the late 1950s and early 60s that Snow achieved the height of his literary powers and influence. Snow was a self-declared progressive; a man of the left who, in spite of a somewhat pessimistic worldview, took an optimistic view of the future – a future underwritten by the possibilities for human progress offered by applied science.

In his Rede Lecture, Snow claimed that the developed nations, including Britain, were entering a new, science-based, industrial era which he characterized as a scientific revolution, qualitatively distinct from the Industrial Revolution which had preceded it and characterized by discontinuous change. Snow saw Britain as singularly ill-equipped to respond to the multitude of changes which this revolution must inevitably bring. Not least he claimed that the country's educational system was not providing the rising generation with the essential knowledge and skills necessary to cope in a world where an understanding and appreciation of science, applied science, technology and engineering would be crucial. This inadequacy, he argued, could be largely attributed to the influence of the intellectuals and specifically what he termed the 'literary intellectuals' with their abiding hostility to industrialization – individuals such as Morris, Pugin and Ruskin, mentioned in Chapter 5. In Snow's view, two cultures had emerged – the one scientific, optimistic and attuned to the modern world; the other literary, pessimistic and hostile to the modern world but nevertheless possessing an inordinate and unjustifiable amount of influence. According to Snow, the huge social changes wrought by the Industrial Revolution had been more or less ignored, or viewed with snobbish disdain, by the prevailing nineteenth-century cultural elite, who nevertheless had gained immense material advantage from the wealth this revolution had generated (Snow 1971). The failure by Britain's cultural elite to engage with the true spirit of the age but, instead, retreat into an illusory world of rural idylls and self-indulgent pessimism had cost the country its industrial edge in the nineteenth century. As Snow saw it, Britain was currently poised to repeat this error unless what we might now term a massive paradigm shift were to occur in social attitudes and educational provision. In Snow's view the key, perhaps the only, means to bring this about, and thereby ensure Britain's survival as a prosperous nation, was to train more scientists, engineers and technicians capable of understanding and promoting the new scientific revolution (Tredell 2012).

In the short term Snow's speech was immensely influential, providing a major ideological underpinning for the Labour Party's victory in the General Election of 1964 under the leadership of Harold Wilson (1916–95). However, although Snow succeeded in projecting a vision of a new Britain based on modernization and progress, he was nevertheless out of tune with the times. In 1958, the economist John Kenneth Galbraith (1908–2006) had published *The Affluent Society* criticising what he saw as the growing phenomena of private affluence and public squalor and dismissing contemporary American consumer culture with a sort of patrician disdain. Galbraith's ideas influenced E. J. Mishan, who published his book *The Costs of Economic Growth* in 1967, challenging the contemporary fixation with economic growth which he saw as merely driving us on to a sort of cultural and environmental wasteland. Mishan pioneered cost–benefit analysis and called for the 'externalities' to be included in the environmental balance sheet when evaluating projects – that is, what might be destroyed as well as what might be gained. The ideas of economists such Galbraith and Mishan, together with the growing influence of Marxist thinkers such as Marcuse (see Chapter 8), served to create an intellectual climate which was increasingly hostile to the ideas put forward by Snow. Further, the attempts by the incoming Labour Government in 1964 to carry through a technological revolution were soon dissipated by trade union resistance and international and domestic economic difficulties – not least an over-valued pound and a rapid decline in international competitiveness. Finally, a bad economic situation was made very much worse by the quadrupling of oil prices which occurred as a result of the Arab oil embargo imposed in the aftermath of the 1973 Arab-Israeli War. Immense inflationary pressures were unleashed in the global economy with dire consequences for Britain which ultimately led to the country having to approach the International Monetary Fund for a US$4 billion loan. In this situation many commentators claimed to discern the beginning of the end for what they saw as the naive belief in progress articulated by Snow (Sheldrake 1991). Among the leading exponents of the alternative view was Fritz Schumacher.

FRITZ SCHUMACHER'S ALTERNATIVE APPROACH

Ernst Friedrich 'Fritz' Schumacher (1911–77) was born in Bonn, Germany. He came to Britain in 1930 as a Rhodes Scholar, studying philosophy, politics and economics at New College, Oxford for two years before spending a year at Columbia University in the USA. By the time he returned to Germany in 1934 the Nazis were in power and he decided to leave, returning to Britain in 1936. When war broke out between Britain and Germany in 1939, Schumacher was interned for three months in a detention camp but spent the bulk of the war working as a farm labourer in Oxfordshire. Combining his agricultural work with research and writing on economic issues, he attended regular meetings with economic and financial experts in Whitehall. From 1946 to 1950 Schumacher was a member of the British section of the Control Commission in Western Germany – a post which involved him in renouncing his German nationality and becoming a naturalized British citizen. His work with the Commission won him the notice of such notables as Sir Stafford Cripps (1889–1952) and Lord Keynes (1883–1946) and led to his appointment to the post of economic advisor to the the recently nationalized National Coal Board (NCB). Schumacher spent 20 years with the NCB, enjoying a particularly close working relationship with the industrialist and former Labour politician Lord 'Alf' Robens (1910–99) who became chairman of the Board in 1961. In 1955 Schumacher accepted an offer from the United Nations to become an economic advisor to the Burmese government. Drawn to mysticism and Eastern religions in general he was particularly attracted to Buddhism. Schumacher later used the insights gathered in Burma to criticise contemporary Western attitudes to work.

During the 1960s and 70s Schumacher became the leading propagandist for the economic and environmental benefits of what he termed intermediate or appropriate technology. In 1966 he became founderchairman of the Intermediate Technology Development Group and travelled widely lecturing on his ideas for smallscale, environmentally friendly technology. He also served as President of the Soil Association (Britain's largest organic farming organization) and as a director of the ScottBader Company (a chemical company which pioneered industrial common ownership). Schumacher retired from the NCB in 1970 and spent his retirement proselytizing for his ideas. Crowded schedules and excessive travel probably hastened his death at the height of his celebrity in 1977. Schumacher published his major work *Small is Beautiful: A Study of Economics as if People Mattered* in 1973. Consisting of 19 chapters and an epilogue it brought together various articles, essays and lectures he had written from the mid1960s to the early 1970s. Central to Schumacher's thinking was the question of size. In this aspect of his work he was strongly influenced by the ideas of the Austrian economist Leopold Kohr (1909–94) and particularly that writer's book *The Breakdown of Nations*, first published in 1957. Kohr developed what he called the size theory of misery, arguing that large-scale social units were the underlying cause of many of the world's problems. Inspired by Kohr, Schumacher mounted a sustained criticism of the propensity of organizations to get ever larger.

According to Schumacher, giant organizations, for all that they might provide a necessary element of stability, were almost bound to result in stultifying bureaucracy, anonymity and sickness. However, notwithstanding his general hostility to largescale organizations, Schumacher was not sufficiently naive to consider that every modern industry could be restructured as a smallscale, facetoface enterprise. His answer to this was the attempt to build smallness within bigness – rather like Mintzberg's adhocracy developing within a machine bureacracy (see Chapter 3). Although Schumacher's strong prejudice was in favour of smallness he nevertheless recognized that the need for operational coherence, order, stability, efficiency and even competitiveness often tended in the direction of excessive organizational size. He therefore saw his role as offering a corrective and sought to establish a compromise between his ideal of smallscale organization and the technological imperatives driving things towards giantism. Schumacher's guiding principles concerning the advantages of smallscale organizations, the disadvantages of giantism, together with the need to develop smallness within bigness, became part of the received management wisdom, prefiguring some of the ideas of Charles Handy and Rosabeth Moss Kanter. Further, the combination in his work of concern for environmental sustainability together with meaningful work influenced a generation of 'alternative' entrepreneurs and retailers such as the entrepreneur and environmental campaigner Anita Roddick (1942–2007), the founder of Body Shop who sought to provide desirable products which were also ethically sourced and produced – which leads us to the topics of corporate social responsibility and business ethics.

THE RISE CORPORATE SOCIAL RESPONSIBILITY AND OF BUSINESS ETHICS

In October 1966 a colliery waste tip from the Merthyr Vale Colliery in South Wales slid down Merthyr Mountain and into the mining village of Aberfan. As a result the local junior school was totally engulfed, killing 109 children and 5 of their teachers. In total over 140 people were killed and the community more or less obliterated. The operators of the colliery, the publicly owned NCB, initially claimed that the slippage was caused by the action of unknown, underground streams. However, as the tribunal of inquiry established to investigate the disaster discovered, the existence of the streams was well known and they were even shown on local maps of the area. The tribunal described the Aberfan disaster as 'a terrifying tale of bungling ineptitude by many men charged with tasks for which they were totally unfitted, of failure to heed clear warnings, and of total lack of direction from above'. Although the chairman of the NCB, Fritz Schumacher's boss Lord Robens, offered to resign his offer was rejected by the government of the day. In the event nobody was prosecuted, dismissed or demoted as a result of the Aberfan disaster. However, the adverse publicity surrounding the deaths of so many children did serve to concentrate the attention of the government on issues of health and safety at work.

Ironically perhaps the committee established in 1969 to investigate and report on the subject was chaired by none other than Lord Robens. The report which ensued led to the enactment of the Health and Safety at Work Act of 1974. In the past 30 years or so a massive change has occurred in attitudes to corporate responsibility. At the time of the Aberfan disaster, public expectations relating to corporate responsibility were very low. Nationalized industries such as coal, gas, electricity, railways, shipbuilding and steel were not exposed to regular criticism in the media for their contributions to environmental degradation. The main focus of criticism was on the treatment of workers and this was articulated through the activities of the various trade unions concerned. In the case of coal mining, the period 1974 to 1984 was the decade of the miners, as the oil shortages generated by the Arab oil embargo briefly enhanced their bargaining power until the time when the discovery and exploitation of new oil supplies reduced it again. Since the 1980s and the rise of the business ethics movement, major corporations have become aware of the need to present themselves as good corporate citizens, which brings us to the subject of business ethics.

Every individual has certain values or standards which guide their conduct. Equally, they have certain expectations regarding the conduct of others. The origin of these values or rules can often be found in the early cultural training of the individual concerned. All the major faith traditions provide guidance on how we should live our lives, i.e. a set of moral principles or ethics. Only in the so-called advanced societies have such matters been detached from religious teaching and become matters of individual conscience. As long ago as 1859 John Stuart Mill, in his essay *On Liberty*, was warning of the tyranny of the majority, arguing for maximum freedom of expression and asserting the claims of liberal individualism. In putting these claims forward, however, Mill conceded that because we live in societies we have certain social responsibilities and a balance must therefore be struck between freedom and security (easier said than done!). Just like individuals, businesses (and all other organizations for that matter) have certain values or rules of conduct without which it is impossible to operate. Also like individuals, businesses exist in specific legal and cultural contexts which increasingly demand they accept their social responsibilities. Since the 1980s there has been a growing interest in business ethics among academics and the business community in Britain. There are specific historical reasons for this, not least a major shift in economic and public policy. For many decades of the twentieth century there was a view (albeit contested) that the state was the appropriate body to moralize business, often through various forms of intervention, including the process of socialization or nationalization. Running in tandem with this view was a substantial degree of hostility to business which can currently be discerned in the hostility to bankers in the aftermath of the financial debacle of 2008.

During the 1980s the retreat of the British government from many aspects of economic activity stimulated an interest in business ethics. As in many other spheres of management thinking the business ethics movement originated in the USA where, as we have seen, a muck-raking tradition exists within the prevailing capitalist system. In societies such as Britain and the USA where consumption has become more significant in the public mind than production, it is important that corporations show that they are people you feel comfortable buying from. However, in spite of raised public awareness and expectations, scandals continue in both the public and private sectors of the economy. In her 2002 Reith Lectures, which she titled *A Question of Trust*, Onora O'Neill highlighted the ways in which an apparent crisis of trust had developed in Britain and outlined the various regulatory efforts which were being been made to remedy this. In particular she emphasized the loss of trust in public servants and the professions which had developed during the 1980s and 90s. She also questioned the value of some of the efforts to improve accountability, suspecting that a culture of 'box ticking' had emerged, serving to alienate those involved rather than improve their behaviour. Of course, a substantial residue of trust must remain in order for society and business to function at all. Nevertheless, since O'Neill gave her lectures there have been a number of scandals perhaps, as has been noted, most notably those associated with the financial crisis of 2008 but also with the press and the National Health Service. All of this has generated a mood of bitterness and hostility such that we currently have an emerging culture of suspicion in which trust continues to be eroded.

CONCLUDING REMARKS

There is no doubt that the general public in the developed world are aware of environmental issues and also expect certain levels of corporate social responsibility. A key factor in the development of environmentalism has been the issue of nuclear power. For many, the matter was settled decisively with the Chernobyl disaster in April 1986 when a nuclear reactor suffered a sudden power output surge, leading to a massive explosion. The resultant radioactive contamination covered a vast geographic area of Belarus, Russia and the Ukraine and led to over 350,000 people being resettled to other areas. Although nuclear power stations continued to be constructed, Britain began to retreat from the industry under pressure from the environmental lobby – in 1997 nuclear power accounted for 26 per cent of electricity generated but by 2004 the figure had fallen to under 16 per cent. During the 2000s the British government shifted its position, adopting a hybrid energy policy advocating renewable energy sources but also accepting the case for a new generation of nuclear power stations to replace the aging structures built in the twentieth century.

In 2011 the Fukushima disaster in Japan once again raised questions regarding the safety of nuclear power and the risks associated with its use. However, economic problems tend to shift the emphasis in public policy from long-term concerns relating to the environment to short-term issues relating to economic growth. The costs and benefits of growth-seeking measures, for example a new London airport or additional runway space at Heathrow, tend to get lost in the clamour for immediate action. In any case the public is bombarded with contradictory messages relating to the best way forward. For example, for many years we were told that nuclear energy was inherently dangerous and that the future is in renewable sources of energy. At the same time we were made aware that supplies of fossil fuels were diminishing and unlikely to meet future demand – an assertion based on the concept of 'peak oil' originally pioneered by the geoscientist M. King Hubbert (1903–89) in 1956. Also, as energy demand from newly developing economies expanded, competition for supplies of scarce resources became a strategic issue with geopolitical consequences and the risk of armed conflict. Underlying this debate is the issue of global warming, still contested but gaining greater traction. James Lovelock, arguably the leading proponent of the inevitability of global warming, or as he calls it heating, has become a surprising advocate for nuclear power (2010). Lovelock is of course aware that his ideas are contentious and likely to alienate many of his associates in the environmental movement. Nevertheless, he argues that science must prevail over sentiment if humanity is to survive.

CHAPTER 10

CONCLUSIONS

In his novella *The World Set Free*, published in 1914, the English science fiction author H. G. Wells (1866–1946) predicted the construction of an atomic bomb. Wells' book was read by the Hungarian physicist Leo Szilard (1898–1964) and, in 1933, inspired his theory of a nuclear chain reaction. Szilard, who was Jewish, was employed as an Instructor of Physics at the University of Berlin but when the Nazis came to power he lost his job and fled to London. According to the well-known story, Szilard was walking through Bloomsbury on the morning of Tuesday, 12 September 1933 and was waiting to cross the road at a set of traffic lights where Southampton Row passes Russell Square quite close to the British Museum. As the traffic lights changed colour and he stepped off of the kerb Szilard experienced a sudden intimation that if the atom were split a nuclear reaction could be set up, energy liberated on an immense scale and an atomic bomb constructed (Smith 2010). In 1938 two German scientists, Otto Hahn (1879–1968) and Fritz Strassmann (1902–80), demonstrated that Szilard's theoretical model of a nuclear chain reaction was technically possible. Meanwhile, in the same year, Szilard moved to the USA, accepting a teaching position at Columbia University alongside the Italian born physicist Enrico Fermi (1901–54). Szilard was aware of the potentially destructive power of an atomic bomb and deeply concerned that, should the Nazis develop such a weapon, they would undoubtedly deploy it with dreadful consequences. Through various intermediaries, Szilard made contact with Albert Einstein and persuaded him to write to US President Franklin D. Roosevelt (1882–1945) warning of the dangers posed to the world should Germany develop nuclear weapons. Roosevelt was convinced by Einstein's letter and responded by establishing a Uranium Committee, thereby commencing the USA's nuclear research programme. In June 1941, Roosevelt signed Executive Order 8807, widening and deepening the programme, establishing the Office of Scientific Research and Development (OSRD) and appointing Vannevar Bush (1890–1974) as its Director. The OSRD became the organization through which the vast bulk of the USA's wartime research was undertaken, including the initiation and early administration of the vast Manhattan Project which ultimately produced the first nuclear bombs that were dropped on Hiroshima and Nagasaki in August 1945 – effectively ending the Second World War (Pascal Zachary 1999).

As Director of OSRD, Vannevar Bush was responsible for co-ordinating the work of some 6,000 scientists involved in the application of science to warfare. In July 1945 published an article in *The Atlantic Monthly* in which he looked forward to the possibilities for science when the war was ended, which he titled 'As we may think'. The tone of the article was optimistic and among the innovations Bush canvassed was a mechanized private file and library which he called the *memex*. Basically, the memex would be a storage device capable of ordering and manipulating vast quantities of data, and it became an actuality with the development of the microprocessor and the extended development of computing. Among the scientists influenced by Bush's article, Douglas Engelbart (1925–2013) was perhaps the most distinguished, having a massive impact on computing and among other innovations inventing the computer mouse. Engelbart studied electrical engineering at Oregon State University and, following service in the US navy during the Second World War, graduated in 1948 and started work at the NACA Ames laboratory (the forerunner of NASA) in San Francisco. Inspired by Bush's article, Engelbart enrolled on a Ph.D. programme at the University of California at Berkeley and later took a research position at the Stanford Research Institute in 1957 where he later led the Augmentation Research Centre (Pascal Zachary 1999). In 1962 he published his seminal work *Augmenting Human Intellect: A Conceptual Framework* and in 1968 conducted what became known as the 'Mother of All Demos' during which Engelbart demonstrated a stream of original innovations including the mouse, multiple windows, hypertext, outline processing and shared-screen teleconferencing, which taken together set the agenda for future developments in personal computing (Moggeridge 2007).

Engelbart was typical of a hugely innovative era in US science and technology. However, in a recent book, Tyler Cowen (2011) has claimed that the pace of innovation in the US and the West generally has slowed drastically to the point where it has all but ceased. Instead of real game-changers, such as the introduction of the automobile, the application of electric power or the development of computing we are merely tweaking the technologies we already have. As he puts it, we have picked all the low-hanging fruit. According to Cowen, even the Internet is of limited significance, its main impact being in the personal domain, basically the shuffling of information and networking. Emerging economies such as China and India, involved in catch-up growth, are currently prospering through technology transfer – borrowing and implementing the best ideas from North America, Europe and Japan – rather than developing new innovations. Meanwhile, living standards in the developed world are declining and disposable incomes being squeezed in the aftermath of the 2008 financial crisis. This is certainly the case in Britain, where for much of the twentieth century the country declined in terms of manufacturing employment and capability, becoming essentially a service-based economy. This decline was not, of course, terminal and pockets of excellence were preserved in aeronautics, pharmaceuticals and the defence industry. In recent times automotive has also enjoyed a modest revival – albeit under foreign ownership – and niche activities such as Formula 1 continue to prosper. The British government, following the lead of the USA, has recently attempted to encourage renewed activity in the manufacturing sector with a view to rebalancing the economy and reducing dependence on financial services. Allied to this ambition is the process of 'inshoring' work – literally repatriating some of those jobs which were lost to low-cost economies as a result of globalization. There is already modest evidence that this process is succeeding as businesses react to the problems generated by extended supply chains, slow reaction times, variable quality and language difficulties. In the case of the USA things have moved rather quicker, prompted by the reduction of energy costs as a result of exploiting shale gas through the controversial process of hydraulic fracturing or 'fracking'. At the time of writing it is becoming evident that Britain possesses substantial reserves of shale gas and it is probably only a matter of time before these reserves are tapped and the country enjoys the benefits of a cheap and secure contribution to its energy requirements. Having said this, the prospect of fracking is already attracting resistance from environmental groups, concerned that the processes involved might taint drinking water supplies and even cause earthquakes. The pros and cons of this debate are beyond the scope of this book. It does, however, raise a serious issue relating to science, technology and innovation.

Scientific and technological progress in the developed economies is increasingly constrained by environmental sensitivities which, however justifiable, inevitably serve to slow the pace of innovation to the point of extinction. To an increasing extent we desire the benefits of science and technology but are also suspicious and risk-averse. This is certainly the case in energy supply where in recent times a marked governmental prejudice in favour of renewables has inhibited essential investment in less environmentally friendly, but generally more efficient, technologies such as clean coal and nuclear. These tensions are not exactly new. Since the 1960s there has been a tendency in the West to weigh both the costs and benefits of scientific endeavour, together with a growing awareness that issues such as global warming are real and not merely environmentalists' spin. Meanwhile, emerging economies, which have so far been less concerned with pollution and more concerned with economic growth, have made rapid material progress. There is evidence, at least in the case of China, that pollution, environmental degradation and the associated health risks are at last causing concern. Self-evidently, no country having once undergone the process of industrialization and urbanization can, so to speak, willingly retrace its steps and return to its pre-industrial, rural origins. Deindustrialization brings many more problems than it solves; we must therefore live with what we have so far created and, in spite of doubts and fears, trust that science will find a way forward.

BIBLIOGRAPHY

Agar, J. 2004. *Constant Touch: A Global History of the Mobile Phone*. Duxford: Icon Books.

Ansoff, I. 1987. *Corporate Strategy*. Harmondsworth: Penguin Books.

Ante, S. 2008. *Creative Capital: Georges Doriot and the Birth of Venture Capital*. Boston, MA: Harvard Business Press.

Ball, P. 2013. *Curiosity: How Science Became Interested in Everything*. London: Vintage Books.

Barnard, C. 1938. *The Functions of the Executive*. Cambridge, MA: Harvard University Press.

Bilby, K. 1986. *The General: David Sarnoff and the Rise of the Communications Industry*. New York: Harper & Row.

Braverman, H. 1974. *Labor and Monopoly Capital: the Degradation of Work in the Twentieth Century*. New York: Monthly Review Press.

Brinkley, D. 2004. *Wheels for the World: Henry Ford, His Company and a Century of Progress*. London: Penguin Books.

Bucci, F. 2002. *Albert Kahn: Architect of Ford*. New York: Princeton Architectural Press.

Burns, J. M. 1978. *Leadership*. New York: Harper & Row.

Carson, R. 1962. *Silent Spring*. Harmondsworth: Penguin Books.

Chandler, A. 1962. *Strategy and Structure: Chapters in the History of the American Industrial Enterprise*. Cambridge, MA: The MIT Press.

Cohn, N. 1969. *Awopbop-aloobop-alopbam-boom: Pop From the Beginning*. London: Pimlico.

Cowen, T. 2011. *The Great Stagnation: How America Ate All the Low-hanging Fruit of Modern History, Got Sick and Will (Eventually) Feel Better*. New York: Dutton.

Davis, S. 1987. *Future Perfect: Tenth Anniversary Edition*. Reading, MA: Addison-Wesley Publishing.

Drucker, P. 1954. *The Practice of Management*. London: Pan Books.

Drucker, P. 1985. *Innovation and Entrepreneurship: Practice and Principles*. New York: Routledge.

Dunne, J. 2012. *The Motor Bike: The Definitive Visual History*. London: Dorling Kindersley.

Fara, P. 2009. *Science: A Four Thousand Year History*. Oxford: Oxford University Press.
Fayol, H. 1916. *General and Industrial Management*. London: Pitman.
Forty, A. 1995 *Objects of Desire: Design and Society since 1750*. London: Thames and Hudson.
Fraser Darling, F. 1969. *Wilderness and Plenty*. London: British Broadcasting Corporation.
Galbraith, J. K. 1967. *The New Industrial State*. New York: The Library of America.
Gerstner, L. 2003. *Who Says Elephants Can't Dance?: Inside IBM's Historic Turnaround*. Hammersmith: Harper Collins.
Gertner, J. 2012. *The Idea Factory: Bell Labs and the Great Age of American Innovation*. New York: Penguin Books.
Gillespie, R. 1993. *Manufacturing Knowledge: A History of the Hawthorne Experiments*. Cambridge: Cambridge University Press.
Gross, D. 1996. *Forbes Greatest Business Stories of All Time*. New York: John Wiley.
Habakkuk, H. 1962. *American and British Technology in the Nineteenth Century*. Cambridge: Cambridge University Press.
Handy, C. 1999. *Understanding Organizations*, 4th edn. London: Penguin Books.
Harnish, V. 2012. *The Greatest Business Decisions of All Time: How Apple, Ford, IBM, Zappos and Others Made Radical Choices that Changed the Course of Business*. New York: Fortune Books.
Hartley, L. P. 1953. *The Go Between*. London: Hamish Hamilton.
Heilbroner, R. 2000. *The Worldly Philosophers*, 7th edn. Harmondsworth: Penguin Books.
Henry, J. 2003. *Knowledge is Power: How Magic, the Government and an Apocalyptic Vision Inspired Bacon to Create Modern Science*. Duxford: Icon Books.
Heskett, J. 1980. *Industrial Design*. London: Thames and Hudson.
Himmelfarb, G. 2008. *The Roads to Modernity: The British, French and American Enlightenments*. London: Vintage Books.
Hodge, B. 2002. *Retrofuturism: The Car Design of J Mays*. New York: Universe.
Holmes, R. 2009. *The Age of Wonder: How the Romantic Generation Discovered Beauty and Terror in Science*. London: Harper Press.
Huff, T. 2011. *Intellectual Curiosity and the Scientific Revolution: A Global Perspective*. Cambridge: Cambridge University Press.
Jacob, M. 1997. *Scientific Culture and the Making of the Industrial West*. Oxford: Oxford University Press.
Jacoby, D. 2009. *Guide to Supply Chain Management*. London: The Economist.
Jardine, L. 1999. *Ingenious Pursuits: Building the Scientific Revolution*. London: Little, Brown and Co.
Kanigel, R. 1997. *The One Best Way: Frederick Winslow Taylor and the Enigma of Efficiency*. London: Little, Brown and Co.

Kanter, R. M. 1983. *The Change Masters: Corporate Entrepreneurs at Work.* London: International Thomson Business Press.

Kanter, R. M. 2001. *Evolve: Succeeding in the Digital Economy of Tomorrow.* Boston, MA: Harvard Business Press.

Klein, N. 2010. *No Logo.* London: Fourth Estate.

Kotler, P. 2000. *Kotler on Marketing: How to Create, Win and Dominate Markets.* New York: The Free Press.

Kressel, H. 2007. *Competing for the Future: How Digital Innovations are Changing the World.* Cambridge: Cambridge University Press.

Kressel, H. 2010. *Investing in Dynamic Markets: Venture Capital in the Digital Age.* Cambridge: Cambridge University Press.

Kuhn, T. 1962. *The Structure of Scientific Revolutions*, 3rd edn. Chicago, IL: University of Chicago Press.

Landes, D. 2003. *The Unbound Prometheus: Technological Change and Industrial Development in Western Europe from 1750 to the Present*, 2nd edn. Cambridge: Cambridge University Press.

Levinson, M. 2006. *The Box: How the Shipping Container Made the World Smaller and the World Economy Bigger.* Princeton, NJ: Princeton University Press.

Liker, J. 2004. *The Toyota Way: 14 Management Principles from the World's Greatest Manufacturer.* New York: McGraw-Hill.

Lovelock, J. 2010. *The Vanishing Face of Gaia: A Final Warning.* London: Penguin Books.

Lucas, J. 1988. *Boeing 747: The First Thirty Years.* London: Browcom.

Magner, L. 1992. *A History of Medicine.* New York: Marcel Dekker.

Magoun, A. 2003. *David Sarnoff Research Center: RCA Labs to Sarnoff Corporation.* Charleston, SC: Arcadia Publishing.

Marcuse, H. 1964. *One-Dimensional Man: Studies in the Ideology ofAdvanced Industrial Society.* London: Routledge.

Mautner, T. 2005. *The Penguin Dictionary of Philosophy.* London: Penguin Books.

McCorduck, P. 2004. *Machines Who Think.* Natick, MA: A. K. Peters Ltd.

McCraw, T. 2000. *American Business, 1920–2000: How it Worked.* Wheeling, IL: Harlan Davidson.

Meikle, J. 2005. *Design in the USA.* Oxford: Oxford University Press.

Mellanby, K. 1967. *Pesticides and Pollution.* London: Collins Fontana New Naturalist.

Mintzberg, H. 1989. *Mintzberg on Management: Inside Our Strange World of Organizations.* New York: The Free Press.

Mishan, E. J. 1969. *The Costs of Economic Growth.* Harmondsworth: Penguin Books.

Moggeridge, W. 2007. *Designing Interactions.* Cambridge, MA: The M.I.T. Press.

Moore, W. 2005. *The Knife Man: Blood, Body-Snatching and the Birth of Modern Surgery.* London: Bantam Books.

Morita, A. 1986. *Made in Japan.* New York: Dutton.

Nasar, S. 2012. *Grand Pursuit: The Story of the People Who Made Modern Economics*. London: Fourth Estate.

Newhouse, J. 2008. *Boeing versus Airbus: the Inside Story of the Greatest International Competition in Business*. New York: Vintage Books.

Ohmae, K. 1982. *The Mind of the Strategist: The Art of Japanese Management*. Harmondsworth: Penguin Books.

Okasha, S. 2002. *Philosophy of Science: A Very Short Introduction*. Oxford: Oxford University Press.

O'Neill, O. 2002. *A Question of Trust*. Cambridge: Cambridge University Press.

Ovenden, M. 2013. *London Underground By Design*. London: Penguin Books.

Packard, V. 1957. *The Hidden Persuaders*. New York: IG Publishing.

Packard, V. 1960. *The Waste Makers*. New York: IG Publishing.

Pascal Zachary, G. 1999. *Endless Frontier: Vannevar Bush, Engineer of the American Century*. Cambridge, MA: The MIT Press.

Pine, J. 1993. *Mass Customization: the New Frontier of Competitiveness*. Boston, MA: Harvard Business School Press.

Popper, K. 1974. *Unended Quest: An Intellectual Autobiography*. New York: Routledge.

Porter, G. 1973. *The Rise of Big Business 1860–1910*. Arlington Heights Ill.:Harlan Davidson.

Rae, J. 1959. *American Automobile Manufacturers: The First Forty Years*. Philadelphia, PA: Chilton.

Reynolds, S. 2011. *Retromania: Pop Culture's Addiction to its Own Past*. London: Faber.

Ries, E. 2011. *The Lean Startup: How Constant Innovation Crates Radically Successful Businesses*. London: Penguin Books.

Riordan, M. and Hoddeson, L. 1998. *Crystal Fire: The Invention of the Transistor and the Birth of the Information Age*. New York: W.W. Norton.

Rogers, E. 2003. *Diffusion of Innovations*, 5th edn. New York: The Free Press.

Schlosser, E. 2004. *Fast Food Nation: What the All-American Meal is doing to the World*. London: Penguin Books.

Schlosser, E and Wilson, C. 2006. *Chew on This: Everything You Don't Want to Know About Fast Food*. London: Penguin Books.

Schumacher, F. 1973. *Small is Beautiful: A Study of Economics as if People Mattered*. London: Vintage.

Schumpeter, J. 1943. *Capitalism, Socialism and Democracy*. London: Routledge.

Shaw, S. 2011. *Airline Marketing and Management*, 7th edn. Farnham: Ashgate.

Sheldrake, J. 1991. *Industrial Relations in Britain 1880–1989*. London: Pinter.

Sheldrake, J. 2003. *Management Theory*, 2nd edn. London: Thomson Learning.

Shenhav, Y. 1999. *Manufacturing Rationality: The Engineering Foundations of the Managerial Revolution*. Oxford: Oxford University Press.

Shurkin, J. 2006. *Broken Genius: The Rise and Fall of William Shockley Creator of the Electronic Age*. New York: Macmillan.

Slade, G. 2006. *Made to Break: Technology and Obsolescence in America.* Cambridge, MA: Harvard University Press.

Sloan, A. 1963. *My Years with General Motors.* London: Sidgwick & Jackson.

Smith, J. 2010. *Fire in the Sky: The Story of the Atomic Bomb.* Bloomington, IN: AuthorHouse.

Snow, C.P. 1971. *Public Affairs.* London: Macmillan.

Tredell, N. 2012. *C. P. Snow: The Dynamics of Hope.* Basingstoke: Palgrave Macmillan.

Tretriack, P. 2005. *Raymond Loewy.* New York: Assouline Publishing.

Trombley, S. 2011. *A Short History of Western Thought.* London: Atlantic Books.

Uglow, J. 2002. *The Lunar Men: The Friends Who Made the Modern World.* London: Faber.

Underhill, P. 1999. *Why We Buy: The Science of Shopping.* New York: Simon and Schuster.

Veblen, T. 1899. *The Theory of the Leisure Class.* Oxford: Oxford University Press.

Weiner, M. 2004. *English Culture and the Decline of the Industrial Spirit 1850–1980.* Cambridge: Cambridge University Press.

Wilson, A. N. 2012. *The Potter's Hand.* London: Atlantic Books.

Womack, J., Jones, D. and Roos, D. 2007. *The Machine That Changed the World: How Lean Production Revolutionized the Global Car Wars.* London: Simon and Schuster.

Woodham, J. 1997. *Twentieth-century Design.* Oxford: Oxford University Press.

Woodham, J. 2004. *A Dictionary of Modern Design.* Oxford: Oxford University Press.

Wren, D. 1994. *The Evolution of Management Thought*, 4th edn. New York: John Wiley.

Wright Mills, C. 1951. *White Collar: The American Middle Classes.* Oxford: Oxford University Press

Wrigley, E. A. 1988. *Continuity, Chance and Change: The Character of the Industrial Revolution in England.* Cambridge: Cambridge University Press.

Wu, T. 2010. *The Master Switch: The Rise and Fall of Information Empires.* London: Atlantic Books.

INDEX